THE NEW WINDMILL SERIES

General Editors: Anne and Ian Serraillier

137

MARY ANNING'S TREASURES

This is the story, based on fact, of a young girl who lived early in the nineteenth century on the Dorset coast. She helped to support her family by collecting fossils from the beach and selling them. Her greatest find, at the age of twelve, was the skeleton of a prehistoric monster, an Ichthyosaurus. The book is rich in fascinating detail and historical background and filled with the suspense and keen pleasure of discovery.

MARY ANNING'S TREASURES

by

HELEN BUSH

Illustrated by Gwyneth Cole

HEINEMANN EDUCATIONAL
BOOKS LIMITED · LONDON

Heinemann Educational Books Ltd
LONDON EDINBURGH MELBOURNE TORONTO
SINGAPORE JOHANNESBURG AUCKLAND
IBADAN HONG KONG NAIROBI

SBN 435 12137 5

First published by Victor Gollancz Ltd 1967
First published in the New Windmill Series 1969

Published by Heinemann Educational Books Ltd
48 Charles Street, London W1X 8AH
Reprinted by photolithography and bound in Great Britain
by Bookprint Limited, Crawley, Sussex

To all my nieces and nephews

CONTENTS

I

A PROMISE

MANY TIMES DURING the night, the raging autumn storm had awakened Mary.

First she had heard the heavy rains beat down. Later strong winds had arisen. They had shaken the little house unmercifully and had rattled the windows furiously. Soon after, she had heard the angry waves tossing the beach pebbles about. Then, as the tide came in, the waves began pounding the sea wall itself, on which their house was perched.

When Mary awoke early in the morning before dawn, the storm was almost over. She raised herself out of bed and peered out of the small dormer window of her tiny attic bedroom. Everything was misty grey except for the white horses that ruffled the surface of Lyme Bay, off the coast of Dorset. The waves had slackened their attack on the sea wall. They no longer flung pebbles high but merely rolled them restlessly backwards and forwards on the beach.

"I hope Father will still take us collecting curiosities on the beach," she said softly to herself. "The storm seems nearly gone."

She loved the stone curiosities that her father brought home from his wanderings on the beach. Delicate starfish and shell-fish, branching sea-lilies, the gracefully coiled ammonites that resembled tightly curled up snakes, and many others—all of cold hard stone.

Her father said that once they had all been alive and lived in the sea. But that was a long time ago. How long nobody knew. And now, like magic, they were all changed to stone. Such a curious thing. No wonder they were called *curiosities*.

When her father had enough curiosities on hand, he put

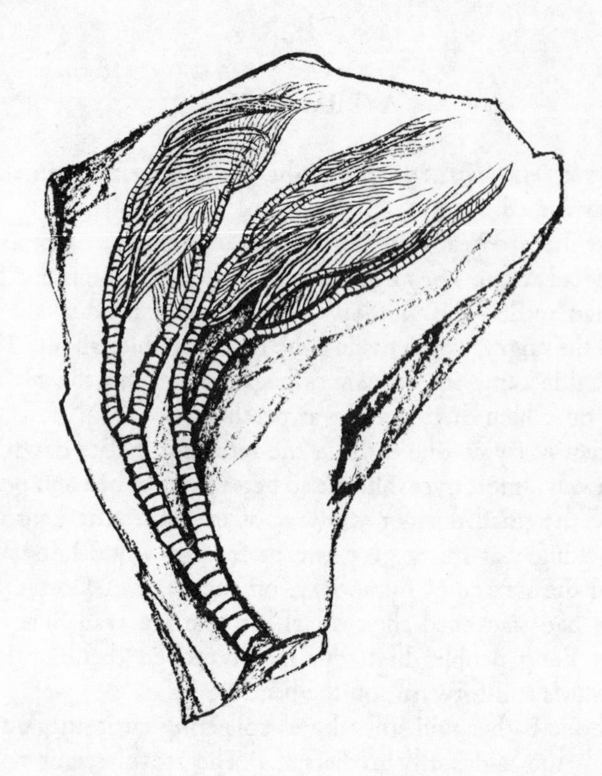

them out for sale on a rickety table outside his workshop. Often he allowed her to arrange them in an attractive manner. How she loved that job! But now they were all sold and he must go collecting again.

And yesterday, for the first time ever, he had said, "Mary girl, if the weather is good tomorrow, I'll take you and

Joseph collecting with me. Then you'll see where I find my best curiosities."

She could hardly wait!

Stepping out on the plaited rush mat beside her bed, she quickly dressed. Her coarse brown homespun dress was well-worn but neat, as were her long black woollen stockings and her high-topped leather boots. In front of a small cracked mirror, she combed her golden curls.

Before she started down the narrow dark stairway, she knocked on her elder brother Joseph's bedroom door, just in case he had not awakened yet. Then she ran lightly down the steps that led into the kitchen.

The flickering flames from the crackling fire in the fireplace lit up the little kitchen so brightly that candles were not needed. The fragrance of wood smoke filled the air.

Her mother was busy stirring the porridge in the black iron pot that hung on a hook above the fire. Her cheeks were rosy from the heat of the fire. Although her fair hair was supposed to be pulled into a tight knot at the top of her head, several strands had escaped and made a soft fringe about her face. She turned to give Mary a quick smile.

Her father, tall and gaunt with curly black hair, stood close to the fire warming himself. He greeted her affectionately, "Good morning, Mary girl."

She ran to the wooden basin that sat on a small bench in the corner. She filled it with water from the jug that stood on the same bench and washed her face and hands. The water was icy cold and she was glad to rub her face and hands hard with the coarse towel to warm them up again.

Seeing only wooden spoons laid out on the table, she reached up to the shelf above and took down four wooden bowls. She placed them neatly beside the spoons. From the cupboard beside the fireplace, she reached for a jug of milk and put it on the table. Then she sat down on the bench by the table to wait for breakfast. Her father soon joined her. Sliding up close to him, she whispered softly, "The storm is nearly gone, Father. Do you think we'll be able to go collecting?"

But her mother had heard, and, before her father could answer, she turned about sharply. "Richard Anning! Surely you aren't going on the beach today to pick up those silly stones."

"Now, Mother," he gentled, "you know they aren't just ordinary stones. They are *curiosities*. When the passengers get off the London coach, they like to buy them for souvenirs. The holiday-makers like them, too. And Mr. Henley—and the Misses Philpots—they count on me to gather curiosities for their collections. And they do bring in a little extra money. Not much, I grant you, but a little."

"Precious little," sniffed Mrs. Anning, "for all the time you spend gathering them."

"Now, Mother," protested Mr. Anning, "you know that I don't neglect my carpenter's shop, but put in a full day's work each day through the week." Mrs. Anning nodded in agreement, but reluctantly.

"And the walk in the fresh air helps my condition." Mrs. Anning nodded again.

At that moment Joseph stumbled sleepily into the kitchen, his mop of black hair tousled and uncombed, his dark homespun jacket and breeches rumpled, his boots unbuttoned. He, too, made his way to the wash basin. Mary and her father grinned as they saw him come wide awake the moment he splashed his freckled face with ice-cold water.

When he had finished washing, he turned to his father and burst out anxiously, "Do you think the storm has gone down enough so that we can go—" Suddenly he saw his mother frowning darkly. He left his sentence unfinished and sheepishly joined Mary and his father on the bench.

"And another thing, Richard Anning," Mrs. Anning spoke

up again, "I don't think you should take the children collecting on the beach. It may lead to them spending all their spare time there like you do.

"We are a poor family. I don't think it is right or proper for you to encourage the children to use their time and energy like this. Collecting curiosities is for gentlemen and ladies of wealth and leisure.

"And with your poor health—you should not be walking so far or carrying such heavy loads. If anything happened to you —" Mrs. Anning did not finish but looked pleadingly at her husband.

"Now, Mother, you worry too much. And it's good for the children to know how and where to collect curiosities. You never know when they may find this knowledge useful."

Mary listened to her parents anxiously. She hoped with all her heart that her mother would not forbid them to go out with their father today. She had often searched the beach close to their town of Lyme Regis, but most of the curiosities there were deeply embedded in stones of great size, far too large to carry home.

The curiosities her father found, however, were smaller and completely free of enclosing rock. They were the ones his customers liked to buy. What fun it must be to collect them! If only her mother liked the curiosities as she and Joseph and her father did!

And today was Sunday, the only day she was free from school. Six days a week she attended the church school, from early in the morning until late in the day. Also, it was the only free time for Joseph. His work at the upholsterer's shop kept him busy for even longer hours, all through the week.

Her mother was still protesting.

"And furthermore, Richard Anning, I don't think it's the proper way to spend Sunday."

"But we'll be back for the eleven o'clock service, Mother."

Then he added teasingly, "That is, if we ever get our breakfast."

Mrs. Anning was startled. She had forgotten that she was making breakfast. Hastily she turned and lifted the porridge pot from its hook, carried it to the table and began ladling the steaming hot porridge into the wooden bowls. Each one helped himself to milk from the jug. Then all was quiet, except for the snapping and crackling of the burning wood in the fireplace, and the sounds of eating as the family heartily enjoyed their good porridge.

When they had finished, Mr. Anning winked at the children and motioned towards the door that led into his workshop. Quickly they rose. Mrs. Anning said nothing. But out of the corner of her eye, Mary saw her frown disapprovingly as they took down their outdoor clothes that hung on wooden pegs near the door. Quietly they slipped out and closed the door gently behind them.

In the little workshop cluttered with lumber and tools, and fragrant with the smell of fresh sawdust, the three soon donned their long cloth coats. It took Mary a moment longer than her men folk because the strings of her sunbonnet had to be tied under her chin. So excited was she, her fingers seemed to be all thumbs.

As Mr. Anning picked up a basket that held hammers and stone-chisels, Mrs. Anning appeared at the kitchen door with a parcel. Without a word, but with a weak smile she handed it to him.

"You're a good wife—you remembered that we'd need a snack," he said gratefully.

When they stepped outside, they found that it was not yet daylight. A damp grey mist shrouded the sleeping village below and above them, for their home was mid-way up the hill. The only person to be seen was the night watchman, just disappearing into a side street at the bottom of the hill.

Then a few sharp barks caught their attention. It was Tray, the fishmonger's little black-and-white dog, coming to visit Mary.

"How on earth did you know Mary would be up and about at this early hour?" asked Mr. Anning scratching Tray's head as the little dog pressed against Mary's legs affectionately.

"Oh, Tray, I'm sorry. I haven't time to play with you now," said Mary as she gave him a few quick loving pats. Then she pointed in the direction of his master's home and off he trotted, apparently satisfied with even a little attention.

Briskly the three set off up the little street, tightly lined on both sides with narrow houses topped with thatched straw roofs. A squeaking sound startled them as they passed one of the houses. But it was only a sign with a painted portrait of the Duke of Monmouth, swinging in the dying wind.

"I think the Duke needs oiling," commented Joseph with a grin.

"Was there ever a *real* Duke of Monmouth?" asked Mary.

"Good gracious yes!" exclaimed her father. Joseph looked at her in surprise.

"He was a pretender to the British throne, on which his uncle James II sat, over a hundred years ago," explained her father.

"Let's see—this is 1810—yes, exactly 125 years ago, he landed with many followers in our town of Lyme Regis. He and his officers made their headquarters in this house and ever since it has been called Monmouth House.

"He hoped to gather more followers to help seize his uncle's throne. Many of the villagers joined him, but they lived to regret it when he was defeated in battle and they were severely punished."

By now they had reached the Parish Church at the top of the hill. Here they turned seaward along a well-worn path close beside the church, that led them to the edge of a jutting cliff. Church Cliff it was called.

Low on the eastern horizon, they could see a faint silver light appearing. It would soon be daylight. The shore of the gently curving Lyme Bay was beginning to emerge. The two humpy hills that stood guard at the end of the bay to the east, Stonebarrow Hill and Golden Cap, loomed up like grey ghosts. Not until the sun was up would they show their true colours—grey rocky fronts and green carpeted backs.

Now the straggly little path descended sharply from the cliff and the three scrambled down, clinging to bushes here and there to keep themselves upright.

A short walk across a pebble beach soon brought them to rippled mud flats, left behind by the tide which was slowly moving out. Here the walking was more comfortable as they set out at a fast pace towards the east.

Next came smooth flat rocks covered with patches of brown seaweed.

"Mind that seaweed, children. It's slippery stuff," cautioned their father. "Better walk slowly and choose your steps carefully."

Soon they came to a golden sand beach that stretched far ahead of them. Ah, thought Mary, this is better. Now we can really hurry. To their left towered the straight-up-and-down face of Black Ven, sheer forbidding cliffs that edged the shore for some distance.

"Keep away from the cliffs, children," warned Mr. Anning. "They aren't as safe and solid as they appear. Especially after storms. The rains loosen the topmost rocks and you never know when, or where, they will fall."

"I was afraid you wouldn't take us collecting when there had been a storm," said Mary.

"Oh?" said her father in surprise. "But that's the best time."

"Is it?" asked Joseph and Mary together. "But why?"

"Wait and see," he said mysteriously. "In the meantime, we'll walk as far from the cliffs as possible—close to the edge of the sea."

As if to emphasize their father's warning, a cluster of stones tumbled noisily from the cliff ahead. Then a larger rockfall crashed down, and, finally, a tremendous slab of limestone thundered down and broke into a thousand pieces below. Dust and fragments flew about wildly for a few seconds.

"We see what you mean, Father," said Joseph wryly, and Mary nodded her head vigorously in agreement.

II

A LESSON IN COLLECTING CURIOSITIES

THE SHEER ROCK cliffs of Black Ven were petering out. Nature was slowly softening the once steep slopes and coating them with a thin layer of soil and grass.

"How much further do we have to go?" asked Mary, impatient to get on with the collecting.

"See that black tongue of mud and rock that has slid down beyond Black Ven?" said her father. Ahead, in a deep crease that furrowed the hill from top to bottom, the children could easily see the dark mud-slide.

"That's where I find my best curiosities. You see, all curiosities are hidden first in the hard rock of the cliffs. Taking them out is dangerous. It is also hard work. But if I wait until the storms loosen the edge of the cliffs and tumble them down—and if I reach the broken rock before the waves wash it away—I often find that many curiosities have fallen out. That's why I look forward to storms along our coast, for they do so much hard work for me.

"By the way," he said, taking two hammers from his basket, and giving one to Mary and one to Joseph, "these will be your hammers from now on."

The children were overjoyed. First they swung them about proudly. Then they began whacking at every rock in sight, for many now littered the beach.

Suddenly Mary gave a cry of delight. She had caught sight of a coiled ammonite of grey limestone. It was entirely free of the rock that had once imprisoned it. Triumphantly she snatched it up and dropped it in her father's basket. She was surprised that he didn't express pleasure at her first find. Instead he gave her a curious little smile.

Soon they reached the mud-slide.

"Why, it's higher than you, Father," exclaimed Joseph.

"See how it's covered with bits of rock? Rocks of all sizes," pointed out Mr. Anning.

All three began to inspect the slide carefully. Suddenly Mary saw another coiled ammonite half-hidden in the mud.

"Oh Father! Joseph!" she shrieked, "I've found another ammonite. A beauty! A perfect beauty!"

She stared at the tightly coiled ammonite in fascination.

"No wonder some people called them snake stones, eh?" remarked Mr. Anning.

About the size of the palm of her hand, each curved section showed up clearly and sharply, from the smallest one in the centre to the largest one at the end. She compared it to the first ammonite that she had picked up. To her surprise, the one from the beach was smooth and worn, the sections just barely visible. She hadn't noticed that it was not even a complete one—that part of it was missing entirely. No wonder her father had not complimented her on finding it.

"It was rolled about so much by the waves that it is both broken and water worn," explained her father. "I'd throw it away, lass. Since we have to carry back our finds so far, we shouldn't burden ourselves with imperfect ones."

Mary tossed it into the sea.

She saw Joseph pounce excitedly on something. "What are these, Father?" he asked eagerly. "These stones that are smooth and long and thin—and pointed at one end."

"Some people call them finger stones. Others call them thunderbolts, for they think that they have fallen from the sky during thunderstorms. I don't believe that. I have often found them when there hasn't been a thunderstorm for weeks. I wouldn't bother picking them up. The coach passengers and the summer visitors wouldn't be interested in them, neither would Mr. Henley or the Misses Philpots."

Mr. Anning and Joseph went on searching. But the slender pointed stones fascinated Mary. So smooth and perfect. Like slate pencils. She couldn't leave them just for the waves to toss about and break into pieces. She slipped them into her coat pocket, unnoticed by the others.

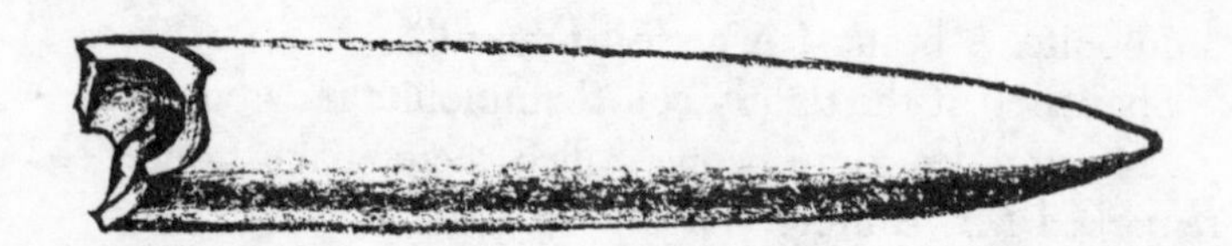

As she continued her examination of the mud-slide, she saw something shining. She picked it out easily with her fingers. It was a small golden ammonite, the size of half-a-crown.

"Oh Father," she cried. "Look what I've found. A *gold* curiosity!"

"Oh no, Mary girl," said her father quickly, before he had even looked at it. "It's *fool's* gold!"

"Fool's gold?"

"It's an iron mineral. It's not *really* like gold. Gold is yellow, you know, not brassy like this," he answered simply. "And besides, gold is not found in our country.

"My customers like these curiosities best of all. Better wrap it up carefully in one of these cloths I brought along."

The children were surprised. They hadn't noticed the cloths in the basket.

"The finest curiosities should always be wrapped to prevent

them from being scratched by the others. Another thing. Those metal ammonites will shine even more if they are rubbed hard with a stiff brush. That is something you and Joseph can do at home," said Mr. Anning.

Suddenly Mary saw something else shining in the mud. "Oh, Father, I've found three more gold—I mean, fool's gold—ammonites. Each is about the size of a sixpence." She was blissfully happy and reluctant to part with them. But finally she wrapped them carefully in a bit of cloth and dropped them in her father's basket.

She saw Joseph frantically searching, anxious to find something special too. When their father reached up high on the slide and pulled out a dark-grey limestone ammonite, the largest one yet, Joseph's gloom deepened.

"Cheer up, son, you never know when your luck will turn. But let's all search more systematically, so that we are not all looking in the same area at the same time—or all missing the same spots."

Mr. Anning picked up four pieces of driftwood lying on the beach and laid them down at equal intervals at the foot of the slide.

"Now, Mary, you search the slide between these two sticks. Joseph, you hunt between the next two sticks, and I'll hunt between the last two. When we have finished, we'll move the sticks on and start all over again."

Soon Joseph gave a shout. "Look what I've found. Two fairy hearts!"

Mary and her father rushed over to admire his finds—two smooth, flattish, heart-shaped sea urchins, each with its five arms delicately outlined on the surface of the stone. Generously Joseph gave one to Mary. She curled her fingers round it. It was just the right size to be hidden in her hand.

She was delighted, then ashamed. She hadn't thought of giving him one of her lovely, golden ammonites. Joseph was always like that, quick to share. And she, if she liked something, was loath to part with it.

Mary's next find was a whole cluster of small stony ammonites still embedded in a slab of rock. Nature had not freed them.

"Sometimes customers like their curiosities better in the rock," said her father, "for the rock sets them off nicely." He held it up at arm's length. "See? It would make a fine ornament on somebody's mantelshelf.

"But you had better break away some of the rock, lass. It will be too heavy to carry as it is."

Mary was just about to strike the slab a strong blow when Joseph grabbed her arm. "Take care, Mary, or you'll spoil your curiosities!"

"Joseph is right, Mary girl," agreed her father. "Maybe we had better have a lesson first in the art of breaking down rock. First, each of you get yourselves an ordinary piece of rock about as long as your foot."

When the children had done this, Mr. Anning took from his basket an assortment of stone chisels. He studied the rock Mary had chosen. "Hmm. Yours is rather fragile, Mary. I'll give you a fine chisel."

With his own chisel, he scratched each of the slabs exactly in the centre. "Now, let's pretend these scratches are curiosities and you are going to break down the rock so that it won't be so heavy to carry. But still you don't want to harm your curiosity.

"Just chip away at the edge carefully with your hammer. Then as you get nearer the centre, use your chisel also. But remember, do it gently!"

Painstakingly both children chipped away at their slabs, gradually reducing the size. But soon Mary became impatient and began working a little faster and more carelessly. To her humiliation, her fragment of rock broke right through the scratch.

Without a word, her father picked up another slab, scratched it in the middle as before and handed it to her. This time she used her chisel and hammer with much greater care. Soon both children had reduced their slabs to handier sizes, but without harming the centres.

With a grin at Mary, her father handed Joseph the curiosity-filled rock that Mary had found. She was glad, for Joseph was sure to be more cautious than she, and the lovely cluster of curiosities would come to no harm.

When Joseph had successfully broken the slab down, Mary wrapped it carefully and stowed it away in the basket.

"Sometimes your rock slab will be too thick," said their father. "In that case, you stand it on end and wedge off a few layers, using both your chisel and hammer." Here he picked up two thick slabs and handed them over to the children to practise on.

After Mary had prised off a layer, her mouth fell open in amazement. "Father! Joseph! Look what fell out. A lot of little ammonites! And you can even see where they have been, for they have left their impression in both sides of the rock."

"You never know where Nature will hide her treasures," observed Mr. Anning.

"Oh, Father, that's a lovely word for curiosities. Treasures!" cried Mary. "That's what it has been like today—hunting for treasures!" Her father smiled at her.

"We mustn't forget to keep an eye on the tide," he reminded them. All three turned towards the sea. It was still far out. They looked back to Lyme Regis. The Cobb was well above the sea. The Cobb was a very old stone wall, over 400 years old, that curved out into the bay in front of the town. It acted as a harbour for boats and sailing ships.

"As long as you can see the Cobb clearly, you have plenty of time for collecting before the tide comes in," advised their father.

Mary was surprised to find that it was full daylight. The chilly, grey mists had disappeared. Sea and sky were a deep blue, while the sun shone brightly. Above the rocky cliffs, a green countryside rolled back to the horizon. She turned towards Stonebarrow Hill and the Golden Cap. No longer were they grey ghosts. A patchwork of soft green fields edged with dark green hedges carpeted their backs; while their steep fronts disclosed their true make-up—layers and layers of grey rock.

Joseph's mind was not on scenery, however. "Let's have lunch. I'm starving," he said.

"Come to think of it, so am I," agreed Mr. Anning. He took from his pocket the simple lunch of bread and cheese and they hurriedly ate it standing up.

Just as they finished, Joseph began to chuckle. Mary looked at him, then followed his eyes. In the distance, jogging along happily towards them was a little bundle of black and white.

"Oh, Tray, you mischief!" cried Mary. "What will your master think?"

All three waited until the little dog caught up with them. He wriggled all over with friendliness, wagged his tail furiously, then jumped up on each of them in turn and tried to kiss them with his pink tongue.

"Well, Tray," said Mr. Anning, "you'll just have to stay and help us collect curiosities."

When they started back to work, Tray was more hindrance than help. He stayed so close, first by one, then another, that he almost tripped them up. Finally when the patience of both Joseph and his father was at an end, Mary pulled the little dog down beside the basket and spoke to him severely.

"Now, Tray, you must stay here and be a good dog." He seemed to realize that he was in disgrace and hung his head in shame. Then he lay down, curled up in a ball and hid his head beneath his paw so that just his bright eyes peeped out. Mary almost burst out laughing but she knew that if she did, he might not obey her.

As Mary began working on the mud-slide again, she saw an unusual stone. It was perfectly round, ridged on the edges, and hollowed out in the middle on both sides. As she studied it

curiously, her father watched her and answered her unspoken question.

"It's a verteberry."

"A verteberry! What's that?"

"Part of the backbone of an animal. Likely a crocodile."

"A crocodile, here?"

"Yes, but a long time ago."

"Why do you think it belonged to a crocodile?" asked Joseph, who had joined them.

"Someone once found part of a jaw, which he said was a crocodile's, although I never saw it myself."

Mary stared intently at the small bone. To think a crocodile had lived here. She thought they only lived far away in warm countries. How strange!

"How did these curiosities come to be hidden in the rocks, Father—and why?" asked Mary.

"Nobody knows, Mary girl. It's a real mystery."

Mary stood lost in thought over the wonder and strangeness of it all.

"Oh, Mary! What difference does it make? They are here for us to pick up. Let's get going," cried out Joseph impatiently.

Joseph was the next to find something special—a slab covered with curving stems, each made up of tiny round plates. Other little plates that had come loose, lay flat side up between the stems. There were little holes in their centres. All were a pinkish colour and showed up strikingly against the grey rock in which they were embedded. Once more, everybody stopped to admire.

"But what are they?" cried Joseph.

"Silly. They are stems of sea-lilies," answered Mary.

"Sometimes you find flowers on the ends of the stems," added their father. "Flowers with many long feathery petals. But they are very scarce."

"Oh, Father, I do wish I knew how all these beautiful curiosities were changed to stone," said Mary wistfully.

"Only God knows, child. It's His secret."

When the sea-lily stems were added to the growing pile of curiosities in the basket, and the three were back at work, Joseph said longingly, "I wish I could find a stone fish."

"They are scarce, too, boy. You could collect for many a day before coming upon one. Lucky the day when you find either a fish or a sea-lily head."

At that moment, Mary gave a shout, "Here's the beginning of your luck, Joseph," and she handed him a small rock fragment showing the impression of fish scales.

"That's a good sign," commented their father. "Maybe we'll find the remainder somewhere—and if not today, then some other day perhaps—or never. That's the way with collecting curiosities. Anything can happen."

Then he turned to look at the tide. The sea was much nearer than before. All three looked back to the town. The sea was higher around the Cobb.

Suddenly Tray barked sharply. They were startled. He had been so quiet, they had forgotten him.

"Tray, I do believe you were watching the tide, too," said Mary as she petted him affectionately. He was so happy to be noticed again, he seemed to wriggle all over, from the end of his nose to the tip of his tail.

In his effort to make friends again with Joseph and Mr. Anning, he once more got in their way and was as much of a nuisance as before.

"Tray!" scolded Mary, "you *must* be a good dog!"

Immediately he looked ashamed and then tried to make himself as small as possible behind the basket.

It was not long after they returned to work that their father gave the signal to stop. "That's all for today, children. No more collecting until next Sunday." Mary looked at the tide mark and with her eye measured the safe margin of beach that still remained. I must always remember this margin, thought Mary.

The basket was so full by this time, that Joseph tucked as many curiosities as possible into his pockets and carried some in his hands. Mary hoped no one would notice her already bulging pockets. The finger stones were her *secret* treasures. She smiled to herself in a most satisfied way.

As Mr. Anning bent to pick up the heavily laden basket, Tray trotted importantly to the front of the little group. And, to their amusement, he led them back along the beach, his long tail pointing straight to the sky.

The shriek of gulls caught Mary's attention as they wheeled about. First the gulls beat their wings hard to rise high, then they held them motionless as they drifted slowly down in great circles.

"Look, Father, aren't they having fun?" said Mary. "Nothing to do but play all day, in the fresh air and sun."

"Well, daughter, I'm not sure about them playing all day—" But Mary interrupted him.

"They are so free. Outdoors all the time. I'd like to be free like that—free to hunt for curiosities whenever I wanted to. School is so dull and gloomy."

"Mary girl, we want you to have as good an education as possible. You know, it isn't every poor girl who gets a chance to go to school."

"I know, Father," admitted Mary, "And I like reading and writing and arithmetic. But such a short time is spent on it. We mostly spin and weave and knit and do the schoolmistress's housework, all cooped up in small dark rooms. Then the other girls are so silly and noisy."

"Yes, I've noticed that you don't have any close friends among your schoolmates, Mary. I'm not sure that's right."

"If I could just be outdoors all the time, like this. That would be better than friends. To smell the fresh salty air, to feel the sun on my face. Lots of space around me, the sea on one side, the hills and downs on the other side."

"Now, Mary, you will have to consider yourself lucky to be collecting once a week and make the best of it," said her father gently. "And maybe you should try harder to make friends. In the meantime we had better hurry along. Joseph and Tray are getting far ahead of us."

Soon they reached the steep cliffs of Black Ven. Both children kept close to the edge of the lapping sea.

"Good," murmured Mr. Anning. "I hope you always remember."

The little town of Lyme Regis gradually loomed up clearer as the three trudged along, each silent with their own thoughts. Mary's thoughts had returned to the curiosities themselves. How could a fragile sea shell turn into stone or metal? And why? And how long would it take? Years? Hundreds of years? It was like a miracle, a stony miracle. If only there was someone to answer her questions.

Suddenly she was startled to find that their father was not with them. She spun round. Not far back, she saw him slumped on the beach, his face as white as paper.

"*Joseph*," she screamed, "look at Father!" Frightened, they raced back.

"Father, what's wrong?" cried Mary. He was breathing with great difficulty but tried to soothe their fears.

"Just resting, Mary, just resting," he answered in laboured gasps.

It was frightening for the children to watch him. It wasn't the first time that they had seen their father ill in this manner, but it never ceased to alarm them. In a short time, however, the colour returned to his face and his breathing became normal. He got up slowly and picked up the basket, but with much effort.

To free one of his hands, Joseph quickly handed some of his curiosities to Mary. Then he took hold of the basket to help ease his father's burden. Purposely the children set a slow pace, giving their father time to recover. They were not surprised when he said pleadingly, "Let's not mention this to anyone, eh, children?"

III

THE HIDDEN FINGER STONES

As they came down the hill towards their little house, Mary saw her mother leave her place at the front window. The door was quickly opened and Mrs. Anning stood there anxiously awaiting her family, her look of deep concern taking in her husband first.

He answered her unspoken question with comforting words, "I feel champion, Wife." The children exchanged a quick glance as their father continued. "It was a good expedition. I think the children know a lot more about how to collect curiosities. With three of us working, we collected far more than I could possibly find alone."

Her mind somewhat eased, Mrs. Anning next turned to her children. Her smile, on seeing their happy faces, faded and changed to a look of exasperation when she looked down at their clothes.

"Curiosities weren't the only thing you brought home," she remarked. "I see plenty of blue lias, too."

For the first time, Mary and Joseph saw that their clothes and shoes were liberally splotched with mud. And to their amazement it *was* blue.

"What did you call it?" asked Mary.

Her father hastened to explain. "Lias. That is what the layers of limestone and shale in our cliffs are called."

"But why?" chorused Mary and Joseph.

"All because people once pronounced the word, layers, as

if it was lias—but into the shop with both of you. We'll scrape off as much of the blue lias as we can. We'll have to hurry for we have barely enough time to get ready for church. When we return, then you can take the curiosities down to the shore and clean them."

They were just about to close the door behind them when a soft little whimpering reminded them of the uninvited guest on their expedition. Quickly Mary turned back.

"Good little Tray, we almost forgot you." She bent down and gave the little dog a hug. Out came a wet tongue to kiss her in return. "Now, away you go to your master."

Satisfied that he had been given a proper farewell, he trotted off obediently. Mary watched him until he turned the corner for his home, before she shut the door.

Joseph and her father had set their burden down and were busily scraping the bluish mud from their boots. Mary had another problem besides cleaning herself. What to do with all the finger stones she had stored in her pockets? When she thought her father and Joseph weren't looking, she hurriedly dumped them behind a pile of lumber in a dark corner of the little workshop. Later she would find another hiding place. Although Mary was certain she had not been seen, she wondered at the sudden twinkle that appeared in her father's eyes.

By the time the church bells began pealing, they were changed into their Sunday best and waiting in the living room for Mrs. Anning to give them a last minute check.

"How unruly your hair is, Joseph." She frowned good-naturedly as she smoothed it with her hand.

"And, Mary, you didn't tie your ribbons at all neatly." She retied the ribbons of Mary's plain brown bonnet that matched her equally plain brown dress.

Then she inspected her husband, who gave her a wink as he waited his turn. She looked him over carefully and fondly, but found only a speck of dust on the shoulder of his homespun suit to flick away. After she had given her own faded, black gown a few pats to smooth away some wrinkles, they left the house to join the throng of villagers flocking up the hill to the Parish Church.

Most of the people, like themselves, were dressed in plain and sober clothes. So it was the few wealthy that drew Mary's attention—the gentlemen with their tall beaver hats and their long coats set off with velvet collars; the ladies with their enormous poke bonnets, gaily decorated with flowers or ribbons or feathers, sometimes with all three, and their billowing gowns of rich silk or velvet almost sweeping the ground. Their children were dressed much like the grown-ups.

"La, Mother, don't they look grand!" murmured Mary as they entered the church.

Mary loved their church. It was very old and full of history. Inside, the beautifully carved pulpit had been given by a mayor of Lyme Regis over 200 years ago. Carved on one of the pillars were the initials of another mayor who had lived 300 years ago. That was when the new part of the church was built. The old part was twice that age. And the town, it was said, was much older again. At least a thousand years old.

Were the curiosities as old as that? Usually Mary followed the service carefully but today she found herself retracing in her mind, every step of their thrilling morning on the beach. She glanced at Joseph. A happy expression covered his face. She suspected that his thoughts, too, were the same as hers.

When they returned home, the rich fragrance of a savoury beef stew met them. Mrs. Anning had left it cooking over the fire. Only once a week could the Annings afford meat and Sunday was the day they chose for this wonderful treat.

"Oh, Mother, it smells so good!" exclaimed Mary sniffing in with deep breaths.

"And we're so hungry!" added Joseph.

"That's right, Mother, we're all as hungry as wolves," added Mr. Anning.

In a very short time the family sat down thankfully to the once-a-week meat and bread soaked in rich brown gravy. They ate slowly, the better to enjoy it and to make it last longer. To end their meal, Mrs. Anning gave them piping hot tea in large pottery mugs.

When the meal was finished, Mr. Anning turned first to his wife. "Mother, that was a fine meal." Next he turned to the children. "Mary and Joseph, away you go and change your clothes. Then you can take the curiosities down to the sea. Better make two journeys. Take brushes, and wash and scrub your curiosities well, for you'll find that blue lias mud is mighty sticky stuff."

Soon the children had changed back into their everyday clothes and were in the workshop emptying the basket of part of its contents.

"You take one side of the basket, Mary, and I'll take the other," ordered Joseph. They carried it out of the shop and round to the sea wall, where a narrow spiral staircase within the wall led them down to the sandy beach. Near the lapping waves, they piled the curiosities. After the second trip, they dipped each curiosity in the salty water, then attacked it with stiff brushes to clear away the sticky mud.

Just then some boys and girls, who had been playing further down the beach, saw them. They raced over.

"What are you doing, Joseph and Mary? Is it a new sort of game?" they asked.

Mary pretended not to see or hear them. But Joseph turned and explained pleasantly, "Oh, we're just cleaning up some curiosities for my father."

"Curiosities? Oh, they're just stones!" said one girl in disgust and as she saw Mary's back still turned, she began to taunt her in a sing-song voice. "Mary likes stones. Mary likes

stones." Then, giggling and whispering together, the children left the two stone-cleaners.

"Mary!" scolded Joseph, "Why were you so unfriendly? No wonder the children tease you."

Mary tossed her head. "I don't care. They're silly and so—so common!"

"And what are you, may I ask?"

" Well, I suppose I am, too, but I don't *feel* common."

"Mary, Mary, quite contrary!" Joseph shook his head in puzzlement. "I still think you should be more friendly to the boys and girls even if you don't like them. You are not Lady Vere de Vere, you know. You are a common girl, too, and you have no right to behave as if you were better than you are."

"But I don't like the things they do. They're silly—and they play silly games!"

"No matter. You should still be friendly." Then both went back to work.

Unnoticed by Joseph, Mary had brought down her finger stones, tucked in her pocket. She washed and dried them, then tucked them away again.

When their task was completed and they had lugged all the curiosities back to the shop and placed them on shelves, Joseph stood back admiring their handiwork. "They do look much more beautiful, with all that gooey mud off them, don't they?"

Mary did not answer. She was wondering where to hide her secret treasures. At that moment, their father entered the shop. He picked up a small stiff brush and began showing Joseph how rubbing the metallic ammonites firmly would make them shine more brightly. Mary looked into the kitchen. Her mother was bending down cleaning out the fireplace.

Tiptoeing softly by her, she slipped into the room at the front of the house. It was a small room, as were all the rooms in their home. It was plainly furnished except for a beautifully carved, heavy oak chest that stood in the corner. She had decided to hide her treasures there, that is, if there were room.

She heaved up the heavy lid and peered in. Although it was well filled with blankets and quilts, she could easily drop her smooth stones into one corner and no one would be the wiser. She emptied her bulging pockets of the stones, closed the lid quietly, tiptoed out again, then back to the shop to join Joseph in polishing the metal ammonites.

"How would you like to arrange the curiosities on the table outside tomorrow, Mary girl?" asked her father.

"Oh, Father, you know I'd love to!"

It was a long tedious week for Mary, waiting for Sunday so that they could go collecting again.

The only two bright spots were when she helped arrange the curiosities for sale and the day she happened to be at home when the London coach rumbled through the village. Usually it came and went while she was still at school. But this day, the coach was late. She was coming home for her midday meal and was greatly surprised to see it rocking down the hill, drawn by four handsome coach-horses, puffing and snorting. The coach passed her and stopped below in front of the Inn.

Mary waited in front of her home to watch the passengers climb out, stiff from their long ride. The men wore great-coats and tall hats. The women were dressed in long close-fitting coats and velvet bonnets. One of the ladies had noticed her father's table in passing, and to Mary's delight, the entire party crossed the street and started the climb up the cobble-stoned hill to see what the table displayed.

"And what have you here, lass?" asked one of the gentlemen in a deep hearty voice.

"Curiosities, sir. My father collects them on the beach," Mary answered politely.

"How interesting!" murmured one of the ladies. Then they touched them.

"They look like sea creatures—but they can't be, for they are stone," exclaimed another.

"Yes, ma'am, they were sea creatures," said Mary shyly.

"But how were they changed to stone?"

"No one knows, ma'am," answered Mary. By this time all the passengers were picking up the curiosities, turning them over and marvelling at their stony beauty, scarcely believing their eyes.

"What wonderful souvenirs they would make!" exclaimed one of the gentlemen.

Mary stepped into the shop and called her father. Still wearing his carpenter's apron and sprinkled with sawdust, Mr. Anning came out to wait on his customers. There were many oh's and ah's as the ladies and gentlemen made their selections and paid Mr. Anning the very modest prices.

Soon, most of the curiosities were sold and the travellers went on their way back to the Inn, chattering with pleasure over their unusual gifts.

"It's a good thing we're planning to collect more curiosities on Sunday, isn't it, Father?" said Mary, looking at all the empty spaces on the table.

"Right you are, my girl," he said as he rumpled her golden curls, then returned to his work.

Saturday night finally came round. Lost in tingling anticipation of the next day, Mary was suddenly shocked into awful awareness. Her mother had opened the oak chest and discovered the finger stones. She was furious.

"Mary Anning! How could you!" Mary cowered under her mother's fury. "Stones in my beautiful chest! The chest that Lady Chester gave my great-grandmother long ago! And muddy stones at that!"

Mary was hurt to hear her lovely curiosities called muddy

stones. "But, Mother," she tried hard to explain, "they weren't muddy. I had cleaned them all nicely."

"You still should not have put them into my beautiful chest without my permission. It is my only treasure."

And the stones are *my* treasures, thought Mary, but she dared not say it aloud. Her mother continued to scold her. Mary remained silent. She could see that Joseph was completely startled by the whole incident. Then he gave her a quick look of sympathy. Good. She had one person on her side.

Just then her father came in from the workshop to see what was amiss. Mrs. Anning lost no time in enlightening him.

He turned to his daughter with a gentle reproach. "Mary girl, you shouldn't have hidden the stones in your mother's prized chest. Why didn't you ask me for a place to put them?"

Mary was silent for a moment. Then she spoke in a low voice. "But you told me not to bring them home."

"Yes, but I didn't realize that you wanted them so badly. Besides I knew you had them."

"You did?" Mary was amazed at first. Then she remembered that sudden twinkle in her father's eyes after she had hidden the finger stones in the workshop. Of course, she should have realized that he had seen her. And she should have known that he would understand.

Mrs. Anning spoke sternly, "I think Mary should be punished by not being taken with you tomorrow."

This was the worst blow of all. She had never wanted to do anything more in all her life, and she had waited six long days. She could see her father's face fall and Joseph's too. She burst into tears.

Immediately her mother softened. "No, that's too harsh. Come and get your stones and your father will find a place for them in his workshop. But, Mary, you must promise me not to *hide* any more of your stones."

Mary stopped sobbing. She saw a look of relief come over

both her father and Joseph. "Oh, Mother, I promise, I promise."

Before the children went to bed, Mary's finger stones were safely stowed away in a drawer in the shop and all was calm in the Anning household.

IV

CAUGHT BY THE TIDE

SUNDAY MORNING FINALLY arrived. Since the tide always went out an hour later each day, they could not set off as early as they had last week. It was not until eleven o'clock that Mary, Joseph and their father met in the workshop to get ready for their expedition. This time each of them was to carry a basket with his own hammer and chisels.

Mrs. Anning stood in the doorway of the kitchen, looking worried and displeased. "Richard, I don't think it's right that you should miss the morning service. You know Mary is supposed to attend every church service because she goes to the church school."

"But Mother, we have to leave now, for the tide is going out. You know the old saying, 'Time and Tide wait for no man,' " he answered patiently. "It won't hurt Mary to miss church once. Besides we'll be back for the evening service."

Mrs. Anning still looked unhappy. "And your health, Richard. Are you sure you are well enough? You look very pale, today."

Mary and Joseph looked quickly at their father. Her mother was right, thought Mary, his face is almost white, not pink like Joseph's or her mother's. If her mother ever knew that he had suffered a sudden attack on their expedition last week!

But he had asked them not to mention it. Should she have told her mother? Which was right—to warn her mother, obey her father—or be selfish and please herself? For there was that to think of, too.

Her father must have been reading her thoughts and been aware of her predicament; for he suddenly ruffled her hair and murmured affectionately, "Mary girl." That decided it. She would not tell, wrong though it might be.

"Mother, you worry too much. I feel fine. You just think about all the curiosities we'll be bringing back—and all the money we'll make from them." He patted her cheek gently as they went out of the door, dressed in their outdoor clothing and carrying their baskets.

"Richard! Richard! You are a stubborn man. You and your stones!" she spoke despairingly but nevertheless smiled at him.

Her family waved to her gaily, then turned and set off up the hill. The streets were deserted, as nearly everybody was in church. Mary suspected that her father had waited until this moment so that they would not be seen by the church-going villagers.

"I wonder if Tray will follow us today?" said Mary hopefully.

"No," answered Joseph firmly.

"But how can you be so sure?" asked Mary.

"Because I asked the fishmonger to be sure he didn't," said Joseph with a grin.

"Joseph! How could you!"

"That's a good idea, son," said Mr. Anning. "You know, Mary girl, your doggy friend wasn't much help to us."

"Oh, Father. You too!" wailed Mary.

The day was bright and sunny. When they reached the edge of Church Cliff, they could see that the gently rolling sea was as blue as the sky. To the east, Stonebarrow Hill and the Golden Cap stood out clear and bright with their patchwork of green fields and hedges.

Hundreds of gulls drifted lazily above the sea, occasionally uttering piercing shrieks.

"Still want to be a gull?" teased Joseph as they began slipping and sliding down the side of steep Church Cliff to the beach below.

"Only if I couldn't look for curiosities," said Mary stubbornly.

As Joseph moved on ahead, Mary spoke softly so that only her father could hear. "It's so good to be out of doors—feeling the warm sun out—and smelling the fresh sea air."

Her father nodded understandingly.

"If I could stay outdoors all the time, I'd never be lonely—I know I wouldn't!" she declared.

"Mary, Mary, you're a lucky girl to be going out even on Sundays."

They crossed the stony beach to the mud left behind by the slow-moving tide, then turned eastward. Joseph was ahead, running carelessly over the flat rocks covered with the slippery brown seaweed. Suddenly they saw him skid and land flat on his back. Mary and her father shook with laughter.

"Wouldn't *you* like to be a gull now?" called out Mary.

Looking foolish, Joseph said nothing in return but hastily jumped up and picked his way more carefully on the treacherous seaweed. Soon they reached the sandy beach. When they came to forbidding Black Ven, all kept well away from its towering cliffs.

To make time go faster as they trudged along, Mary coaxed her father to tell them again of the famous writer who had often visited their village.

"Oh, you mean Miss Jane Austen," smiled Mr. Anning. "Well, once upon a time, when you were a very little girl, Miss Austen came to my shop. She was tall and graceful and beautifully dressed in a billowing gown of blue muslin and lace,

with a matching poke bonnet and parasol." Mary liked this part best.

"She had a small jewel box with a broken lid and she asked me if I could repair it. While she waited, she admired the curiosities on the table outside. And she talked to you, even though you were very shy and said hardly a word back to her."

"And why was she in our town?" put in Mary, knowing full well.

"She had come to enjoy the sea air, as did many grand people every summer.

"Even though Jane Austen was only a young woman, she had written several books and was famous all over Britain, maybe farther away than that, for all I know. She walked so often on the Cobb, that people were certain that she must be planning to write about it in one of her books sometime."

"It must be wonderful to be famous—and known all over the country," said Mary longingly.

"Don't be so envious," teased Joseph. "Maybe you'll be famous some day."

"For what?" asked Mary half seriously.

"Let's see—mm-mm. Maybe for finding curiosities," he teased.

Then he raced down the beach, Mary after him in pretended anger. When they were both out of breath, they plopped themselves down on the sand to wait for their father to catch up.

When Mr. Anning joined them, they hurried to the dark mud-slide.

"A person would never know that it held hidden treasure," said Mary thoughtfully. She was the first to see an ammonite among the litter of stones on the beach. As she bent down to pick it up, she gave a cry of surprise.

"What is it, Mary girl?" asked her father.

"It's the same worn and broken ammonite that I picked up last week and then threw into the sea," she said in amazement.

"The tide brought it in again," explained her father. Exasperated, Mary walked to the very edge of the lapping waves and with all her strength, flung it far out, much farther than before. That would be the last of *that* curiosity!

Before they fell to work, their father made an unexpected announcement.

"By the way, children, I am going to let you watch the tide and decide when we should return home. Making that decision at the right time is a most important part of curiosity collecting, and I want to see if you are equal to the responsibility."

The children looked seaward. Yes, the tide was far out. Looking back, they could see that the Cobb stood high above the water. All was well for the present.

The day's search began.

As they had done last week, they first marked off individual areas with pieces of driftwood, so that there would be no overlapping in their search. It was a satisfying morning as the pile of curiosities in their baskets grew steadily larger. As before, the ammonites were the most numerous. One group displayed different colours from the week before. They varied from a soft, creamy yellow to a dark chocolate brown. Even a single ammonite sometimes showed both colours, one fading into the other.

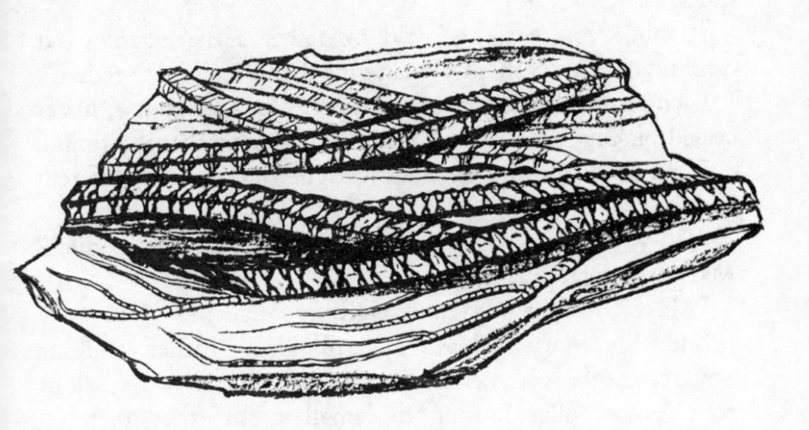

"That creamy colour is like honey," observed Mary, "and the dark brown reminds me of chocolate. I know what. I'll call them honey and chocolate ammonites!"

"Good description," approved her father.

But all Joseph said was, "Girls!"

Then Mary gave a shout. "Look what I've found!"

"Look what *I've* found!" echoed Joseph. When they held up their finds and realized that they had both picked up exactly the same curiosity, heart-shaped sea urchins, they broke into laughter.

The next triumphant shout came from Mary. "Father! Joseph! I've found a little slab of sea-lily stems. The stems are all twined about, and they are of shiny fool's gold. It's beautiful enough to be made into a brooch," and she held it against her dress. After it was duly admired, they returned to work.

All was quiet for a time until Joseph exclaimed excitedly, "Look! I've found a fish curiosity. I do believe it's part of the one we found last week." Once more the three were in a huddle, considering the small fragment covered with stony scales.

"I think you're right," said his father as he examined it carefully. "If so, perhaps we can glue the two pieces together."

Each resumed searching once more. Mary, of course, never ceased picking up each and every finger stone that she found.

From time to time, all found verteberries of many different sizes.

"Do you suppose they all belonged to the same crocodile?" Mary wondered out loud.

"Maybe. Maybe not," answered her father.

Suddenly Mary had a wonderful idea. "Father, couldn't we arrange the verteberries on the table as if they *did* belong to one crocodile? Putting the smallest ones first, then the larger ones?"

"Good idea, Mary girl," answered her father.

Then Mary came upon a new curiosity. It was about as long as her thumb, half curled around, and with fine lines radiating from one end to the other. "What is this lovely curiosity, Father?" she asked excitedly.

"That? It's a devil's toenail," he answered casually. As Mary looked startled, he hastened to explain further. "It was really a cockle of some kind, but people thought it resembled the devil's toenail. So they call it by that name." Then he chuckled, "Just how they knew what the devil's toenail looked like, I can't imagine."

"Oh Father, I wish I knew how these animals changed into stone and metal. And how long ago they lived. It's all so mysterious! Did people live then, Father?"

"I have never found any human bones. But, I wouldn't bother your pretty head about it, Mary girl. All these long-ago remains are God's secret—and yet they do say there are new men, called scientists, who are studying the earth, trying to unlock its secrets."

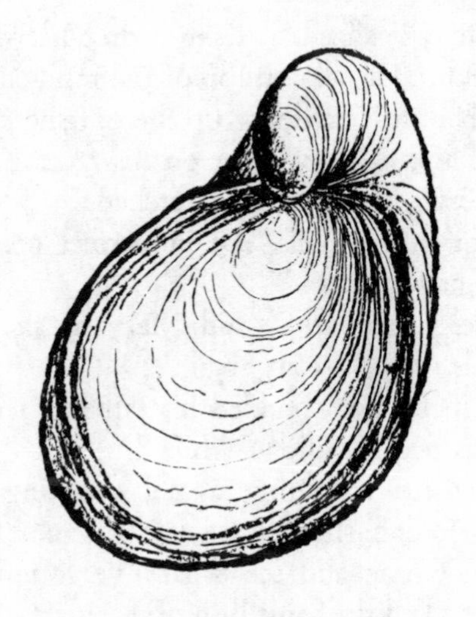

"That would be so exciting. I wish I could hunt for curiosities *all* the time, instead of being shut away in a little, dark, cold room every day," Mary replied.

"That would be a strange occupation for a girl, or even for a woman, you know. The Misses Philpots like curiosities but they don't go collecting on the beach."

"Why not?"

"Well, it—it just isn't proper." Her father seemed to be

searching for a good reason. "It's a man's work and poor man's work at that. Even Mr. Henley, the Lord of the Manor, doesn't do his own collecting."

"Well, I think both the Misses Philpots and Mr. Henley are silly then," said Mary stubbornly. "Look what fun they are missing."

Mr. Anning smiled at her impatience.

At that moment, Mary remembered to look at the tide. It was coming in but was not too close yet. Still more time for collecting.

Mary suddenly realized that she hadn't heard from Joseph for some minutes. He was on top of the mud-slide almost out of sight. "What are you doing up there?" she called.

He glanced up, a happy grin on his face. Then he began scrambling down, his hand outstretched.

"Look, Mary. Six snails. All in perfect condition. Here, you can have half of them."

"How lovely they are!" said Mary as she admired the graceful spirals of each stony snail.

"You're a lucky boy," added his father. "I haven't found any as good as that for a long while."

All returned to collecting again, working quickly and silently until Joseph shouted excitedly again from the top of the slide, "Come and see what I've found—the biggest ammonite yet. And the most beautiful. Hurry!"

Mary and her father scrambled up the slide on all fours.

"Oh, Joseph! A giant honey and chocolate ammonite," Mary exclaimed puffing from her climb.

"It's as large as a coach wheel, I declare!" Mr. Anning shook his head in wonder. The huge ammonite was neatly cut in half, as if done by man. The inside was much showier than the outside. The curving chambers, spiralled gracefully from the centre; while, within the hollow chambers, sparkled small clusters of creamy crystals. It was breathtakingly beautiful, and the collectors simply stood before it in dumb admiration.

Mary was the first to break the spell. Gently she ran her hand over both the smooth part and the sharp pointed crystals. She looked up at her father. He was still silent, studying the beautiful curiosity thoughtfully.

"You know, children, this has given me an idea. Many of the ammonites we have been gathering may be as beautiful inside as this one. If Mother Nature can cut them in half, so can we—but we'll use a saw. It will be long tedious work sawing the ammonites, but it can be done."

"I'm certain that people would like to buy them for ornaments—for the mantels on their fireplaces perhaps."

"That's a lovely idea," exclaimed Mary, while Joseph's eyes lit up with enthusiasm.

"Is there any way we could take this one home?" begged Mary. "It's too lovely just to be left here in the mud, with no one but us to see it."

"I don't think so, daughter. It's far too large to transport so far. We'll just have to leave it here. Some day the rains and storms will carry it down the mud-slide. Then the sea will gradually destroy it. It seems a pity, too."

This reminded Mary to look at the tide. To her horror, it was well in. "Joseph! The beach is nearly all under water." She looked westward. "And the sea is washing against Black Ven! There's no path there at all!"

Joseph was as shocked as she was. They looked at their father. But all he said, as he quickly picked up his basket of curiosities, was "There's nothing for it but to climb the hill," and he set out quickly along the narrow path of beach not yet reached by the tide. Mary and Joseph, thoroughly ashamed at forgetting to watch the sea more carefully, grabbed up their baskets and kept close to their father's heels.

After a long minute of silence, he continued. "There's a

part of Black Ven where it is still possible to climb. It is very steep, but we can do it. It will be slow going and we'll likely be too late for evening service."

The children knew all the meanings held in the last statement. Their mother would be angry. She might even forbid them to go on any more collecting trips. And what about their father? Could he stand such a hard climb?

Their collecting had been almost too successful for they were all so heavily burdened with their stony treasures, it was not easy to move fast. As for Mary, her heavy basket kept bumping against her bulging pockets, laden with her prized finger stones.

The remaining strip of beach did not quite extend to the spot where they could start climbing and for the last few yards, they waded through the sea.

"Mother really will be angry now," wailed Mary. "Wet feet, blue lias mud all over me, and missing church."

"That's not all. Wait and see what we look like after climbing Black Ven," warned her father. Mary looked uneasily at the frowning hill, high, very high above them. At least, at this point, it did not show rock but was covered with thin soil and grass. If they had been caught at the foot of the straight up and down cliffs—she decided not to think about that.

They began the long, hard climb. While the green grass that clothed the slopes gave them a soft appearance, it was really a slippery and tangled mass that made their climb all the more difficult. Furthermore, thistles generously sprinkled the grass and scratched their hands and legs. Their load of stones became heavier and heavier. The bright sun shone down on them without mercy.

Each time Mary looked up, the top of Black Ven seemed no nearer. Her arms ached and her hands were soon cut and bleeding, from the thistles and from clinging to the tough grass. There was no chance of resting, for it wasn't possible to get a firm foothold on the slippery grass. If she stopped, she began sliding down. There was only one thing to do, keep moving up, inch by inch.

She looked over at Joseph. His hands were in the same state as hers. He, too, appeared to be finding the climb just as difficult as she did. She turned to the other side but her father was not there. She twisted round and looked down. He was a few yards below, looking very pale. He was obviously breathing with difficulty and his movements were laboured. How terrible if he were to have one of his fainting spells! She groaned. If only she and Joseph had watched the tide more carefully.

Joseph heard her groan and glanced up questioningly. She motioned with her head towards their father. Alarm spread over Joseph's face as soon as he looked down. He stopped clinging to the hillside and let himself slide downhill until he reached his father. Mary followed suit. Each took some of the curiosities from their father's basket to lighten his load. Then, somehow—later she couldn't have said exactly how—they both pushed and pulled their father, as well as themselves and their burdens.

In the distance, they heard the church bell ringing for the evening service. Mary looked at Joseph. Anyone who was allowed to go to the church school was supposed to attend all church services on Sunday. Her mother would not be the only one who was angry. The wealthy ladies who supported the school would not be pleased either.

Tugging and heaving, they struggled along, inching their way upwards. When it seemed as if they had been climbing for hours and that their agonizing climb would never end, Mary and Joseph suddenly found there was no more hill left. They had finally reached the top!

All three collapsed flat against the ground. They rested their weary arms and legs and let their pounding hearts slow down.

Once their strength had returned and their tiredness had gradually oozed away, the children began to perk up. They looked at their father. He was still flat on the ground, utterly exhausted. They did not let him see that they were ready to go but stayed perfectly still and waited for him to recover fully. When he eventually began to stir about a bit, Mary spoke up softly, "Father, did you know the tide was in before we did?"

"Yes, child."

"Did you want to teach us a lesson?" He nodded.

"But, Father!" protested Joseph, "It has been harder on you than on us."

"Yes, son. Sometimes children—and grown-ups, too, unfortunately—have to learn things the hard way. I won't always be with you, you know, and I wanted you to see how absolutely necessary it is to watch the tide at all times. It can be your friend, but it also can be your deadly enemy."

"What did you mean, Father, when you said that you would not always be with us?" asked Mary in sudden alarm.

He hesitated several seconds before answering. "When I think I can trust you, naturally I will let you go collecting alone."

A chill came over her. Was that what he really meant?

At dusk, a sorry looking, bedraggled trio trudged heavily down the hill towards home. But at least no one saw them; for the villagers were again at church, as they had been in the morning when the three had set out.

Mrs. Anning stood at the open doorway, waiting for them. There was no smile of welcome on her face. First, she looked at the blue lias mud and grass stains that covered them from head to foot. Then she looked at her husband's face. "Richard! You're not well!"

Mr. Anning's hand went limp and his basket fell to the ground with a thud. His body began to slump forward and Mrs. Anning quickly took hold of one arm to support him.

"Joseph, go for the doctor as fast as you can," she ordered. "Your father is very ill."

V

MARY SELLS A CURIOSITY

IT WAS A sad and lonely Mary who wandered aimlessly along the beach, well beyond the village. When her father had died two months ago, it was as if the sun and moon had disappeared from the sky, for her world now seemed dark and empty.

Never to hear her father say affectionately, "Mary girl" again? She couldn't believe it. Yet she had to believe it.

There was no sound of hammer and saw in the little workshop. It was strewn with cobwebs and had become a dingy and dismal place. And there would always be the empty place at the kitchen table.

Mary's mother was also sunk in grief and moved about listlessly, not seeing or caring about anything. She had not even seemed to notice when Mary continued to stay away from school. When the schoolmistress and the ladies who sponsored the school came to complain about her absence, her mother said nothing, just looked at them blankly.

At one time, the most wonderful thing in the world would have been to be free of school with its long hours of eternal sewing and knitting and weaving. Now she was free. But it wasn't wonderful, for her father was gone, and all the free time seemed to stretch out endlessly ahead of her.

Mary's one comfort was Tray. He trotted up the hill several times a day on friendly visits. This day he had carried a basket in his mouth, and tried to coax her away to the special mud-slide. She'd laughed in spite of her sadness. "Tray, you darling, I do believe you want me to go collecting curiosities."

Then her voice caught with sorrow. "It's no use collecting any more, Tray dear, for all the curiosities we gathered with Father are covered with dust and cobwebs—like everything else in the workshop." Tray seemed to understand, for he turned and set off for home with his basket, leaving Mary alone with her thoughts.

Mary still felt terribly guilty. She was certain that their strenuous climb on the last collecting expedition must have caused her father's death. No wonder her mother had tried to keep him at home.

But her mother had tried quickly to allay her fears.

"No, dear, your father has been ill for some time. There really was no hope for him—even though I did hope. While I didn't approve of his going so far along the beach, nor of taking you with him, it gave him great pleasure," she had said sadly.

Their only source of income now was Joseph's pay from his work in the upholsterer's shop. But he was only a boy and he was paid a very small sum. They could no longer afford meat, even once a week. They were lucky to have bread and cheese twice a day. If only I were a boy, thought Mary, I could go to work, too. But I'm not.

Just then she realized that someone was talking to her, a lady, a strange lady. She had thought that she was alone on the beach.

"I am sorry, ma'am, were you speaking to me?"

"Would you sell me that curiosity in your hand, little girl?" asked the lady. "I haven't seen one like that all the afternoon."

Mary was dumbfounded. She looked at her hand. Sure enough, she was holding an ammonite, although a badly waterworn one. She had been so carried away by her thoughts she hadn't realized that she had seen it and picked it up.

Apparently the lady thought that her hesitation meant she was reluctant to sell the curiosity. "I'll give you half a crown for it, little girl," she coaxed. "I always like to have a souvenir stone from each beach that I walk on, and I have seen nothing today as interesting as the one you are carrying."

As if in a dream, Mary handed her the ammonite and took the money in return. A *whole* half crown. And for a poor curiosity at that.

Her dark, gloomy thoughts fled. Bright, shiny ones replaced them. Why couldn't she sell curiosities as her father had done? Now that she had left school, she would have plenty of time to collect. She would be the first lady to collect and sell curiosities. Mary suddenly felt very proud.

With a happy smile on her face, Mary turned back towards the town. Running and skipping by turns, she soon reached the sea wall. As she climbed the steep stairs that spiralled up inside the wall, she began to have second thoughts. What would her mother say? Would she even be interested? She walked slowly home and opened the door. Her mother, sitting in the darkest corner of the little front room, did not even look up.

"Mother, see what I have!" she said and held out the half crown.

Slowly her mother looked up at Mary with dull disinterested eyes. Then just as slowly, she looked at her out-stretched hand. Suddenly her eyes came to life.

"Mary, where did you get that money?" she cried sharply.

"I was walking on the beach with a curiosity in my hand and a lady asked to buy it."

Mrs. Anning was astonished.

"Do you see what that means, Mother? It means that I can sell curiosities as Father did!"

Mrs. Anning was silent a moment, then spoke as if without hope, "Do you really think you could, daughter?"

"Oh, yes, Mother!" cried Mary, happy and relieved to have her mother's attention after so many days of being ignored. "I could sell them from the table outside the shop, the same as Father did. When the ones that we have are gone, Joseph and I could go collecting more. We know where the best ones are —for Father showed us."

Mrs. Anning was caught up in Mary's enthusiasm for a moment. Her eyes lit up and then they dulled again. "I don't know, Mary. I don't know. . . . Let's wait and see what Joseph thinks."

Quietly Mary slipped out to the shop. First she wiped the dust and cobwebs off the window panes so that the light could shine in more brightly. She cleared the tools and sawdust off her father's work table. Then she took the curiosities out of the baskets, placed them on the table and began to sort them.

When she came to the fragment of fish, she remembered the other one that they had brought back on their first expedition. She searched the shop until she found it. When the two fragments fitted together perfectly, a cry of joy escaped from her lips, "Father was right, they are part of the same fish . . . but there is still a piece missing. Wouldn't it be wonderful if I could find it some day? Just in case I do, I'll put these pieces aside."

Then she polished all the metal curiosities until they shone.

As she stood admiring her handiwork, she was well pleased with herself. At this moment, Joseph entered. He saw the curiosities neatly stored on their father's work table. "Oh, oh. And what might this be about?"

Excitedly she told him of her surprise sale to the strange lady on the beach. "And so I think you and I should be able to sell curiosities, just as Father did."

Joseph's eyes brightened. "What a wonderful idea, Mary. You know, I didn't see how we could manage much longer on my small pay. But this might help. Let's discuss it with

Mother." They ran into the kitchen where their mother was setting out the bread and cheese and making tea.

Since her father's death, Mary had found the evening supper a sad and quiet meal, but tonight was different. Such fun it was to be discussing plans for selling the curiosities. A faint smile even hovered over her mother's face, the first for a long while.

Finally, half to herself and half to her children, she said thoughtfully, "I wonder if this is why your father persisted in taking you collecting on the beach, in spite of my worry and in spite of his health?"

VI

CURIOSITIES FOR SALE

"JOSEPH, WILL YOU help me carry the table outside before you go to work?"

"Certainly," said Joseph good-naturedly. They had finished breakfast, and Mary, who had been awake long before dawn, was terribly eager to begin her first day of selling curiosities. "Here, Mary, you take one end and I'll take the other." In less than ten steps the table was set out in the sun. "I've a feeling this will be a big day for you. Remember how I once said you might become famous because of the curiosities?"

"Joseph, don't be silly. How could I possibly become famous just for that?"

"You wait and see," he answered with a knowing grin on his face.

"Joseph, you are a tease! But it is exciting to think that I can spend my days collecting and selling curiosities, instead of being cooped up in that dark little class-room all day," said Mary with a sigh of relief. Then she hurried back into the shop.

Joseph waved goodbye and good luck to her as he set out for work, but she was so intent on arranging the curiosities on the table, she did not hear him.

First she placed the verteberries in order of their size across the middle of the table, so that they looked like the backbone of one animal. Then she made a spiral of the ammonites, large ones on the outside, smaller ones inside. She made another spiral with finger stones.

In the remaining space on the table, she arranged some sea urchins, snails, devil's toenails and sea-lily stems. Last of all she set up a new and fresh sign that she had made the night before, CURIOSITIES FOR SALE. Then she waited.

It seemed as if almost everyone in the village passed by that morning. Each stopped to take a quick look at her table and give her a few words of encouragement.

"Never heard of a girl selling curiosities before," said Mr. Griffin, the thatcher, who thatched many of the roofs in the town, "but I wish you luck, Mary."

"Your father would be proud of you," said Mr. Drury, the butcher, his big white apron covering him from his chin to his boot tops.

"What a fine collection you have, Mary." This was Mr. Fair the wheelwright. After her father's workshop she liked his next best, for it, too, always smelled of freshly cut wood. "I hope you have many customers."

"A queer thing for a girl to do," commented one of the housewives out shopping for food, "but it should be a help to your mother."

The doctor in his tall black silk hat stopped to congratulate her on her new business, and so did the Vicar and the chimney sweep and the lamplighter. But the muffin man did more than speak kind words. He generously gave her a muffin.

Once she heard snuffling sounds. It was Tray. He nuzzled close to her, quietly and gently. How strange, she thought. Usually he was excited when he greeted her. She bent to give him a tender hug, then quickly stood up straight. After all, she was in business now. She must always be ready for customers. Tray seemed to understand for he curled up under the table, well out of the way.

Later some of the school children came by. Mary's smile faded. As she expected, they began to laugh and poke fun at her unmercifully.

"Stone girl! Stone girl! Mary is a stone girl!" they chanted in high-pitched voices, dancing backwards and forwards in front of her. Mary was terribly embarrassed. Furthermore, she was afraid that they might drive away customers.

Suddenly there was a fierce barking. It was Tray defending her. The children were startled at first, then they laughed at the angry little dog. But they soon left. Tray retired under the table again, growling softly to himself. Mary's smile returned. Good dog, Tray. He at least, was a true friend.

But then, perhaps Joseph was right. Perhaps it was her own fault that the children teased her. Maybe she was too unfriendly. While she was still mulling over her problem, she saw the Misses Philpots coming up the hill.

Mary stood expectantly. They had been her father's best customers, along with Mr. Henley, the Lord of the Manor. They were fair pink-cheeked ladies, elegantly dressed in full-skirted blue dresses and frilly white bonnets. It was the older one, Miss Annie, who was the most ardent collector. Since it was not considered proper for ladies to do their own collecting on the beach, she had always been dependent on Mr. Anning to supply her with curiosities.

"Oh, Mary," exclaimed Miss Annie warmly, "I'm so pleased that you are collecting curiosities as your father did."

When she saw the profusion of finger stones, she gave a cry of delight. "Aren't they lovely? So smooth and slender. And pointed. I wonder why your father didn't collect them?"

But before Mary could answer her, she exclaimed, "Oh, I must have *all* of them."

"What on earth for?" asked her sister.

"Never mind. I've suddenly had a wonderful idea—but I'll wait until I see if it is successful, before I tell you about it."

Mary was as bewildered as Miss Annie's sister. What could she be planning to do with so many finger stones?

"Mary, please save all your finger stones for us, from now on. I'll need them all.

"And I'll take two snails . . . and some of the ammonites, especially your nice polished ones . . . and this rock covered with small ammonites. That will make a most handsome ornament on our mantelpiece."

That's what Father said, too, thought Mary to herself.

"Oh, I must have this brassy fragment of sea-lily stems. It would be lovely—just lovely—for a brooch," and she held it against her dress, just as Mary had done herself when she had first found it. Miss Annie was a most satisfying customer!

Miss Annie paid Mary for all her purchases, then tucked them away in her shopping basket. The ladies swept away, their handsome dresses almost touching the ground. As soon as they were out of sight, Mary raced triumphantly into the house.

"Mother! What do you think? All the finger stones are sold—the ones that you—" She paused in embarrassment.

"You mean the ones I scolded you about?" said her mother ruefully as she finished the sentence for her. "And the ones your father advised you not to collect?"

Mary nodded. Then she showed her mother the money she had received from Miss Philpots. "And she wants more finger stones. All I collect."

As she returned to her table, she heard her mother murmur, "A person never knows."

Mary took up her position beside the table again. Then she noticed that Tray was still underneath, huddled in a little ball. She pulled him out, gave him a pat and tried to send him home.

"Your Master will be looking for you," she explained.

Tray refused to budge.

"Whatever is wrong with you, Tray? You have always obeyed me before. You know your master will be worried about you if you stay away too long."

"Got yourself a new dog, eh?" boomed out a voice suddenly.

Startled, Mary looked up and saw Mr. Henley, a stout, red-faced man who was Lord of the Manor. The Manor, with its own mill and blacksmith shop and dairy and everything else that a big farm needed, was situated on the top of the hill above the town. In his spare time, Mr. Henley was a collector of curiosities for a small museum in London. Like Miss Annie Philpots, he preferred that someone else should do his collecting.

"He's not my dog, sir," explained Mary. "He belongs to the fishmonger. But he won't go home and I don't understand."

Mr. Henley's face suddenly became serious. He bent down and gave Tray a gentle pat. Then he spoke sadly, "He has no master any longer, Mary. The fishmonger died last night."

"Oh, poor Tray. Now I understand," and Mary knelt down and hugged him tightly. He whimpered to her softly as if he understood that she was sympathizing with him over the loss of his master. Then she gave him a final caress and stood up to wait on Mr. Henley. She was hoping he would find something interesting on her table.

"So you are carrying on in your father's footsteps, eh?" Mr. Henley said as he began examining each and every item on the table. "Good for you. I can see several things here that I'd like—the sea urchins and the snails . . . and these ammonites.

"Good gracious! I've never seen so many vertebrae at one time."

At the unfamiliar word, Mary pricked up her ears. "What did you call them, sir?"

"Vertebrae," he repeated, his eyes twinkling. "Have you been calling them verteberries?"

Mary nodded.

"The villagers have always called them that," he explained, "but it is not the proper name at all, just a nickname. One bone of a backbone is called a vertebra, several bones are called vertebrae."

"Did these verte-vertebrae," Mary stumbled over the new word, "did they belong to crocodiles, sir?"

"Perhaps. I saw part of a jaw once and it looked as if it might have belonged to a crocodile. But we'll never know for sure until we find more of the skeleton. In the meantime, I'll take all the vertebrae you have."

Mr. Henley paid for his curiosities, put the small ones in his numerous pockets and carried the larger ones in his hands. After he had gone a few steps, he stopped and called back, "If you ever find the skeleton of that crocodile—or whatever it is—let me know first, eh, lass?"

"Oh yes, sir, I will, sir," answered Mary eagerly.

She had barely put the money in her pocket when she saw and heard the coach from London rumbling down the hill. Just as the times before, the fine ladies and gentlemen discovered the little shop and the table full of strange and lovely curiosities. And just as before, it did not take long for these people to buy up every single curiosity and to depart well pleased with themselves.

But no one was better pleased than Mary. Excitedly she ran into the house to give her mother all the money that she had made that day. Her mother could scarcely believe that the curiosities had sold so quickly.

"I think we'll buy a little meat for this Sunday to celebrate your going into business," she said.

But now I haven't any more curiosities to sell, Mary told herself—and it's so long to wait until Sunday for Joseph to go with me.

Supper that night bubbled with excitement. Mary recounted her day for Joseph's benefit.

And Joseph had good news, too. One of the town's wealthy ladies had asked him if he thought his mother could do weekly washings for her and some of her friends. Mrs.

Anning was pleased and relieved to know that there was also a way for her to help.

"I wish tomorrow was Sunday and you could come collecting with me," said Mary to Joseph.

At that moment, little whimpering sounds were heard outside. "It's Tray," explained Mary. "His master died last night and he's been here all day. May I let him in for a little while, Mother? He's so lonely."

"Ye-es," said her mother reluctantly, "but if you make him too welcome, he'll want to stay with us. And you know we can't afford to feed another mouth."

Mary opened the door to a dejected little Tray. When he saw her, his dejection quickly changed to delight and he bounded in joyously. Mary welcomed him with many loving caresses. Then he ran to Joseph, who tussled him playfully. Mary noticed with amusement that Tray avoided their mother. Perhaps he sensed she did not approve of him.

Suddenly Joseph stopped stock still. "Mary, I have had an idea. You needn't wait for me to go collecting. Take Tray with you."

Both children looked at Mrs. Anning to see her reaction to this proposal. She did not answer immediately. When she did speak, she spoke slowly as if thinking out loud.

"If Mary is going to sell curiosities steadily, she will certainly need to go collecting more than once a week—but she's too young and there are too many dangers for her to go alone. Tray would be company for her—and if perhaps there were an accident, she would be able to send him back for help."

"Oh, yes, Mother," interrupted Mary excitedly. "He has always obeyed my commands until today. But of course, he had a good reason."

"And Mary, if you find more curiosities than you can carry home," put in Joseph, "just pile them up in a safe place and on Sunday I can go with you to pick them up."

Both children knelt down to hug and congratulate Tray. "Good old Tray!" said Joseph. "You have saved the day."

"And now I can go collecting curiosities any time I want to," exclaimed Mary blissfully.

"Oh?" laughed Joseph. "Even when the tide is high, or there are storms?"

"We-ell, no," admitted Mary sheepishly.

A much friendlier Mrs. Anning now bent down to pat the little black-and-white dog. "Well, Tray, we won't have to worry about feeding you, now you'll be earning your keep."

Happily, Mary made up a bed for Tray in the corner of the workshop. Then she climbed the two flights of stairs to her own little attic room to sleep and dream about her new venture.

VII

A SUCCESSFUL COLLECTING TRIP

IT WAS NEARLY dark when Mary and Tray returned from their first collecting trip. Joseph stood in the doorway of the workshop anxiously awaiting them.

"Well, Mary, you look as if you had a good day."

"Oh Joseph, it was a lovely day! And I was so lucky!"

Tray joined in with a chorus of sharp "wuffs."

"So you want me to ask how your day was, eh?" Joseph grinned as he turned to give Tray a welcome pat. Tray wagged his tail furiously in thanks. Then unexpectedly, he began to limp over to his bed in the corner of the workshop.

"Tray is limping. He must have had a hard day," observed Joseph.

"I doubt it," laughed Mary as she carefully unpacked the basket. "I've been so busy collecting, I haven't had much time to talk to him. He limps to get attention—and the rascal doesn't always limp on the same foot, either."

"Tray, you *are* a rascal!" Joseph shook his head in pretended exasperation. Tray drooped. He curled himself in a ball and hid his head, all except his eyes which looked out reproachfully. When the children just laughed at him, he hid his eyes, too.

Then Joseph turned eagerly to Mary. "Now, Mary, let's see your treasures."

"Just wait until you see my *special* treasures."

Quickly Mary began pulling out a small wrapped bundle.

"Here's my first special curiosity—a beautiful sea-lily head. Remember how Father said they were terribly scarce? See what look like long feathery petals? Isn't it a beauty?

"And mind you, Joseph, I didn't know exactly what I had until I dipped it into the sea and washed the mud off. Such a surprise," and she held the slab at arm's length the better to admire it.

Joseph, being a boy, didn't like to appear unduly excited, but Mary knew by his shining eyes that he really was as thrilled as she at her unexpected find.

"And another surprise, Joseph. You know that old, worn and broken ammonite that I threw into the sea both times that we were out with Father? I saw it again. For the third time the tide had brought it back.

"So this time, I picked it up and dashed it against a boulder to break it up. And what do you think? When it broke, out fell a lot of little ammonites, all of fool's gold. It was just like finding hidden treasure. No wonder the sea had thrown that old ammonite back so often!" and she pulled the shining little ammonites out of the basket to show him.

"Well, Father said the sea could be a friend, as well as an enemy," remarked Joseph.

"And look at *this* treasure. I think it's a sea urchin, but it's different from any that Father ever found. As round as a little ball, but covered with these darling little bumps."

"Humph! They look like warts to me," said Joseph a little scornfully.

"Oh, Joseph, how could you. Warts are ugly. These aren't. See how the little bumps form a lovely design all over the sea urchin?"

"We-ell, maybe you're right," he said reluctantly, then added hastily, "Oh, I almost forgot. Mother's keeping supper for us. I'm starving—you must be, too."

"Yes, I am." Mary had almost forgotten her hunger in the

excitement of showing off her special finds. "We'll look at the rest after supper," she promised.

Tray had come to life at the mention of supper and all three hurried into the kitchen. The fire burned brightly and cast flickering shadows over the room. The tea kettle was boiling furiously as it hung on the hook over the fire. Mrs. Anning, knitting by the fireplace, jumped up with a welcoming smile.

Quickly she made the tea, motioned Tray to his dish filled with scraps, then joined the children at the table lit by a single candle.

"And did you have a good day, daughter?"

"A wonderful day, Mother."

"I hope you didn't try to bring too heavy a burden with you."

"Oh no. I had far more curiosities than I could possibly carry. I piled them by a big boulder so that we would be certain to find them again, and I was also careful to put them well above high water mark.

"Just as I finished, the nicest thing happened—I found the loveliest curiosity of the whole day. A small golden snail, in shale as black as coal."

"Golden?" Joseph raised his eyebrows sceptically.

"Well, perhaps not *real* gold," admitted Mary, "but it had a golden colour, rather than brassy."

"Didn't you bring it home?"

"No. I had forgotten to take a fine chisel with me. The snail was so delicate-looking and the chisel I had brought was so coarse and heavy, I was afraid I might spoil it getting it out of the rock. So I thought it would be better to leave it until I returned with the proper chisel."

"We-ell, I don't know about that." Joseph shook his head doubtfully.

"Oh, I know exactly where it is, Joseph. Right near the large boulder where I piled the other curiosities. Just wait until you see it!

"Then I saw something else of gold," Mary looked at her brother impishly.

"Now Mary, how you do exaggerate!" scolded Joseph.

"It was the setting sun shining on the Golden Cap," she explained quickly. "Now I know how the Golden Cap got its name. It was a dazzling sight. And that was the end of my day —a golden snail and a Golden Cap." She stared off into space dreamily until she realized that Joseph and her mother were looking at her indulgently.

"Have you enough curiosities to keep you going until we go for the rest of them on Sunday?" asked Joseph.

"I think so. I brought all the finger stones for Miss Annie Philpots. I have lots of small ammonites, different kinds of sea urchins, several snails and devil's toenails, and a few verteberr—no, vertebrae."

"Vertebrae?" questioned Joseph.

"Mr. Henley said that was the proper name for verteberries."

"Oh!" exclaimed Joseph in some surprise.

"And how did Tray behave?" asked her mother.

At the mention of his name, Tray thumped his tail.

"We-ell, he was good company for me, but I think he expected me to be better company for him," said Mary.

"Mary says he limps to attract attention," added Joseph.

At these words, Tray hid his head underneath his paws.

"I do declare, I think he understands every word we say," said Mrs. Anning as she began clearing the table.

"And he sniffed seaweed," went on Mary, "and dead fish. He chased any bird that tried to land on the beach—that is, when he wasn't getting under my feet and tripping me up. But I was glad he was there." Then she added in a low voice, "It didn't seem right without Father. I missed him so."

A sadness came over them, while tears filled Mrs. Anning's

eyes. Suddenly Tray scrambled to his feet, trotted over and pressed against each of them in turn.

"Thank you, Tray," said Mary gratefully. Joseph and her mother nodded in agreement. Soon all were smiling again as they showered the little dog with pats and kind words.

Mary awoke in the middle of the night, dismayed to hear the sound of a violent storm. Strong winds shook the little house perched high on the sea-wall, while angry waves pounded the sea-wall below. For several days and nights, the storm raged.

Disappointed at not being able to return to her collecting, Mary cleaned the curiosities that she had brought back and polished the metal ones. Each day she put some curiosities on the table which stood just inside the door out of the wind and rain. But she made no sales.

The coach passengers did not show the slightest interest in her wares. Neither did Mr. Henley or the Misses Philpots appear, although Mary knew that when Miss Annie did come, she would be taking the finger stones at least.

Only the village girls and boys paused as they passed by to tease the "stone girl" and laugh at her odd stones. When loyal little Tray barked angrily at them, they laughed all the more.

Mary complained about the children annoying her, but Joseph scolded her soundly, "If only you would smile at them, Mary, the children might, perhaps, stop and look at your curiosities, instead of laughing at them."

"Oh no. I am sure they wouldn't care two pins about them," cried Mary. "They just want to be cruel."

She worried over her lack of sales, but Joseph and her mother warned her that she would have to expect days like that.

"Just wait until the summer visitors start coming," consoled Joseph, "then your business will be better."

Mary shook her head. The future looked very black to her.

VIII

MARY FINDS A FISH

BY SUNDAY, THE storm had dwindled away.

Joseph accompanied Mary and Tray to help bring back the heavier curiosities that Mary had set aside. Mary had not forgotten to bring an extra-fine chisel to chip out her golden snail. She could hardly wait to get at it. She smiled in happy anticipation as they hurried along the coast.

As the end of the mud-slide came into sight, Mary was surprised to see that it was much flatter. She was puzzled. Oh! Oh! It was the storm. Either the high waves or the heavy rains, or both, had washed much of it away. Then a sinking feeling came over her, as she looked in vain for the boulder near which she had piled her curiosities.

"Oh, Joseph," wailed Mary, "I think the storm has carried away my curiosities. I had placed them well beyond the reach of high tide, but I forgot that a storm could raise the sea even higher. Now surely the boulder wasn't carried away—it was too huge—the largest one here," and with her arms curved, she showed Joseph how large it had been.

After considerable searching, with Tray close at his heels, Joseph found it, but at some distance from where Mary had remembered seeing it, and nearer the sea. As for her precious pile of curiosities and her beautiful golden snail, there was not a trace. Only a thick layer of the sticky blue lias mud plastered the whole area.

"Joseph, my golden snail is gone!" Mary was near tears.

"You were right, I should have tried to work it out with the chisel I had. But I was so sure that it would be here next time I came."

Tray sniffed around the mud vigorously, he too seemed to be searching for her lost treasures. Mary laughed in spite of herself. "Tray, you are a dear."

Tray barked in answer to her as he busily kept on sniffing every inch of the muddy expanse.

"I think his barking means always be sure you have all your tools when you go collecting," explained Joseph with a chuckle.

"Don't worry. I won't forget again," said Mary firmly.

A few minutes later, Joseph called out in a mysterious voice, "Mary, come and help me roll this boulder over."

Her dark despair turned to curiosity as she ran to assist him. When they finally succeeded in pushing the boulder over, her hands flew up in astonishment. "It's that huge honey and chocolate ammonite you found the last time we were collecting with Father. But all the hollows are filled with mud."

"I had a hunch it was the same one," admitted Joseph. "We'll leave it overturned so that at high tide the waves can wash out the mud. Unless there's another bad storm, I don't think it will be carried out to sea. It's too near the high tide mark and too heavy."

"Oh, Joseph, wouldn't it be wonderful if we could take it back to the shop? If it were sitting in the window, it would attract people."

"That's what I was thinking, too," agreed Joseph.

"But how can we possibly get it home?"

"I'm sure we can find a way," answered Joseph thoughtfully, "I'll think of something—and now let's look below the mud-slide and closer to the sea and see if by some chance, some of your curiosities haven't been scattered along the beach."

Carefully and systematically, Mary and Joseph searched the

beach. As Joseph had thought, the sea had failed to carry away some of Mary's heaviest curiosities, especially the larger slabs ornamented with ammonites, which she had left for him to break down. With shouts of joy, Mary reclaimed them. Then she stood by while Joseph expertly reduced their size.

Of her beautiful golden snail, however, there was not a sign.

Later on she came upon a single vertebra, nearly as large as Joseph's cap. The more she thought about it, the more shocked she was. "Joseph, what a huge crocodile it must have been! If this is just one part of its backbone, I don't know whether I'd want to find the whole skeleton or not. It might be frightening."

Joseph laughed. "Come now, Mary. Surely not. Think how exciting it would be, especially as no one else has ever seen a whole one."

Later Mary saw her brother puzzling over a large water-worn boulder. She joined him. "I don't see anything special about it," she commented after a quick glance at it.

"I do," declared Joseph. "See here—and here. Just a faint outline of a few verte—what does Mr. Henley call them?"

"Vertebrae," answered Mary.

"—with the ribs attached to them." He traced them out with his finger. "I think this is more of your crocodile. However, they are so much a part of the rock, that I don't think they will come out separately. We'll just have to leave them."

"Joseph, do you suppose we'll ever find the *whole* skeleton?"

"I am sure we will," he said calmly. "It's just a matter of time."

"Do you ever wonder how these animals and shellfish were changed into stone?" Mary asked as she stood staring at the large vertebrae and ribs.

"Never think about it."

"Joseph!" Mary was shocked. He was a wonderful brother, but at moments like this he could be exasperating.

After checking on the progress of the tide, they left the beach and began searching on the mud-slide. So much of it had been carried away, that entirely new rocks and curiosities were uncovered. There were more than enough curiosities to make up for the loss of the others and their baskets were fast filling up.

Finally Mary's boots became so heavy with mud, she went to the water's edge to wash them off. Suddenly she shouted, "Joseph, come quickly!"

Joseph looked up in alarm, until she added, "A *whole* fish!"

Just under the water, in the flat shale, lay the bony remains of a small slender fish, pressed into the rock.

"I'll have to work fast to get this out, for the tide has already started to come in," muttered Joseph as he hurriedly slid off his boots and stockings. Mary shed hers also, the better to keep an eye on proceedings. Luckily the fish was in a thin slab of shale. Using Mary's fine chisel and his own hammer, working gently and carefully, Joseph was able to prise out the thin slab that held the fish skeleton, all in one piece.

As he triumphantly lifted it from under the water, both he and Mary were startled to hear Tray barking excitedly. So absorbed had they been, they had completely forgotten him. He apparently had been watching them with deep interest, but from the dry beach. He seemed to be as pleased as they were with their successful mission.

They turned back to admire their find.

"What a beauty!" said Mary wonderingly, as she felt it with her finger tips, "The bones show so clearly."

She held it up at arm's length, and studied it. Suddenly she had a wonderful idea. "Joseph, if you could saw this slab square, it could be used as a picture, to hang on the wall."

"You're right," agreed Joseph heartily, "and I think I could do it, if I am very careful—that is, if we don't drown first," he laughed, looking down at the water. It was nearly up to their knees.

They raced back to the dry beach, wrapped their precious fish carefully in cloths and placed it on top of Joseph's basket. They pulled on their stockings and boots, and with Tray at their heels, set off for home.

"Another wonderful treasure hunt," sighed Mary with deep pleasure, then added regretfully, "If only I could have found my golden snail. No wonder Father warned us that the tide was both an enemy and a friend. How right he was."

With their heavy loads, they could not return home as fast as they had come. Because of their slower speed, Mary had time to notice other things, especially the many discarded shells, scattered on the beach. Some were of snails and cockles and sea urchins. Most were broken.

How frail they are, thought Mary, while the ones in my basket are changed to hard stone. Why? How?

And the dead fish thrown up on the beach. The waves would soon break them into countless bits and pieces. Yet, in her basket was a complete fish skeleton, firmly embedded in stone. It had long withstood the sea's attack.

Why were some of God's creatures preserved in stone, while others were being dashed into nothingness?

And if crocodiles and ammonites lived here in the past, why weren't they living here now? Crocodiles still lived in other parts of the world. About the ammonites, she didn't know.

The more she thought about these things, the more confused she became. She would have loved to discuss them with Joseph, but he thought she was silly to be so curious.

Mary was so wrapped up in her thoughts that the time sped by quickly. When the church bells began to ring above her head, she could scarcely believe that they had reached Church Cliff.

Soon, brother and sister were home, had changed into their

Sunday clothes and were accompanying their mother to church. Of course, Tray tagged along, but just to the church steps where he curled up in a corner to await their return.

The rich organ music pealed forth as they entered the church and found their pew. Tired but happy, Mary sank into the seat, letting her eyes roam around to the favourite corners of her beloved church. Then as everyone knelt for prayers, she added her own special thanks for a good day's collecting—even if she had lost her golden snail.

IX

MORE SPECIAL TREASURES

REMEMBERING THAT THEIR father had intended to saw the "honey and chocolate" ammonites in half, Joseph and Mary consulted Mr. Forbes, the local stonemason.

Mr. Forbes was a plump, grey-haired man. With a fine chisel and hammer, he was carving a flowery design on a handsome red granite gravestone. The stone dust that flew up from his hammering and chiselling covered him from head to foot.

As soon as he saw the children, he stopped work. "And what can I do for you?" he said pleasantly.

"We'd like to saw in half some of the softer ammonites we collect, Mr. Forbes. Could you tell us the kind of saw we should use?" asked Joseph.

Mr. Forbes put down his tools and wiped the dust from his face before he answered. "You use a special saw. Like the one over there on my work-table. It is a narrow iron band held rigid in a wooden frame. The band doesn't have teeth like a wood-cutting saw. Instead, as you push the band backwards and forwards on the stone, you keep wet sand in its path and the sand really does the cutting.

"Cutting stone is tedious work and takes lots of patience. It would be too hard for Mary to do. But she could put the final polish on the sawn surface of the ammonites."

"How could I do that, Mr. Forbes?" asked Mary.

"You'll need a slab of limestone first. Next a handful of

fine sand on top of the slab and add a little water. Then rub the sawn surface of the ammonite round and round in the wet sand. Keep adding water to the sand so that it is always moist. Eventually the surface of the ammonite will be ground down smoothly."

"That should be easy to do," said Mary.

"Yes, but it's tedious work, just as Joseph's sawing will be," warned Mr. Forbes. "And that isn't all. You now want a glassy polish. To do this, you rub the ground surface of your ammonite with wet leather covered with fine chalk or lime dust."

Mr. Forbes paused a moment. "Tell you what—I'll ask the ironmonger to make you a saw and then I'll come over and help you to start. I'll bring a supply of fine sand, chalk and a piece of leather also. In the meantime, you get the limestone slab for polishing, also a small barrel."

When the brother and sister looked surprised at the mention of a barrel, he explained further. "The barrel will hold the water for your sawing. It is placed a little above where you will be sawing. The barrel will need a small hole in the bottom with a loose peg in it. This will allow the water to trickle out slowly and keep the sand always wet."

They could scarcely wait to start on their new project.

When the saw was made up, Mr. Forbes brought it over one evening. Mary quickly lit candles in the shop. Mr. Forbes helped Joseph to set up the barrel of water above a small work-bench. Then he chose a small ammonite for Joseph's first attempt at sawing and anchored it firmly on the table. He put a little pile of extremely fine sand close by. Next he wriggled the peg in the barrel above so that water began to trickle down on the top of the ammonite. As Joseph drew the saw backwards and forwards, Mr. Forbes kept adding a little sand directly under the dripping water and in the path of the saw.

As Mr. Forbes had warned them, sawing stone was tedious work. It was also messy, and Mary soon placed a bucket under the bench to catch the dripping water.

When the small ammonite was finally sawn through and the two halves fell apart, it was a rewarding moment for the children—for the gracefully curved chambers showed up much more handsomely inside than on the outside.

Now it was Mary's turn. Mr. Forbes placed a little sand on the limestone slab that they had brought up from the beach,

then added a little water. Round and round, Mary rubbed one of the sawn halves of the ammonite in the wet sand, while Mr. Forbes kept the sand constantly wet by adding water every now and then. Eventually the surface was ground down smoothly.

Next he dipped a piece of leather in water to wet it, then placed a handful of chalk dust on it and gave it to Mary to rub on the ground-down surface. Soon a glassy polish developed.

"Oh, it's lovely, Mr. Forbes," sighed Mary. "I thought the

ammonites were beautiful just as we found them on the beach—but they are even more beautiful like this."

And so brother and sister started in the business of cutting and polishing, as well as collecting.

The next night, before they began work, Mary remembered the fish curiosity that she wanted made into a wall picture. She found it and placed it on the work-table.

"Joseph, let's cut this so I can use it on the wall as a picture."

Joseph thoughtfully studied the piece of flat rock with the bony remains of the little fish in the middle. He found a ruler and measured both the fish and the stone.

"The fish is about four inches long. Now, if I cut it in a ten-inch square—ten inches wide and ten inches long—it should set off the fish rather nicely."

Joseph placed the flat slab on the sawing bench. Since the rock was thin, it did not take long to square it off with the saw.

"Won't it be lovely against the wall!" cried Mary.

"We'll see." Joseph placed it on the top shelf and leaned it against the wall. Then he stood back and considered it. "Yes, you're right, Mary. It's just like a wall plaque and yet it's also a curiosity. Very suitable for your shop."

The weeks took on a set pattern now.

On the three days that the London coach came through the town, Mary remained at home in the hope of selling some of her wares to the passengers. During her spare time she polished. On the other days, if the weather was suitable, she and Tray combed the beach for more curiosities. On Sundays, Joseph joined them and helped bring back any curiosities that had piled up during the week; his evenings were spent sawing.

One evening as they worked together in the shop, Joseph paused a moment in his sawing to rest his arms. Mary noticed him looking at her curiosities piled everywhere in the little shop—on shelves, on the work-table and on the floor.

"I really have a good stock of curiosities, haven't I?" she asked.

"Well, you'll need them when the holiday-makers start coming. What do you keep in the drawers?"

"I keep my finger stones in the top drawer for Miss Annie Philpots," she explained, adding impatiently, "I do wish I knew what she is doing with them.

"The next drawer holds my very special curiosities—the ones I don't think I'll ever part with. In the bottom drawer are my fish curiosities, although I haven't many yet. And that far corner I call my crocodile corner, but there isn't much there either," she admitted wistfully.

Then Mary noticed Joseph pausing in his work to sniff the air.

"Why are you doing that?" she asked.

"I just realized that we can't smell sawdust in here any more. What with my sawing and your polishing, there's only a mud smell. After all, it's no longer a carpenter's shop, but a curiosity shop."

This reminded them of their father, and both faces saddened.

Now that the stormy season was nearing an end, not so many curiosities were turning up; and Mary's basket was not so heavily laden when she returned home from her collecting trips. From time to time, however, she still made an occasional exciting find so that her pile of special treasures gradually grew larger.

One day she had found entombed in the flat-lying rock on the beach, three large vertebrae with long curving ribs attached. She had patiently chipped away, trying to free them from their rocky bed. Before she had accomplished much, the tide had come in. Reluctantly she had to abandon her work.

The next time she returned, she was surprised to find that the churning sea, rolling sand and pebbles backwards and forwards at high tide, had freed the bones considerably more. Again she set to work with her chisels and hammer. But once more she had not finished before the tide had crept in and had begun to cover them.

This time she realized that the sea was her friend and that it would take over where she had left off.

Twice more she returned before she—and the sea—had fully freed the three vertebrae and their ribs. As she stood back proudly admiring her find, she was unexpectedly shocked to see Tray pounce on the bones and start gnawing them.

"Tray, stop that! How *dare* you chew on my crocodile bone!" Mary lashed out at the little dog so angrily that he slunk away with his tail between his legs.

Then she relented. "I'm sorry, Tray, of course you didn't know these bones were any different from other bones. Forgive me."

Tray returned joyously to accept her apologies, but kept well away from the bones.

Another day she noticed something new and strange embedded in the sand on the beach. It looked like the back of a turtle. It was made up of numerous five-sided plates, all fitted together like a mosaic. She touched it. It was of bone. It must be a curiosity. When she tried to pick it up, she realized that there was more of it hidden under the sand. She brushed the sand away until it was all uncovered.

There appeared to be seven long toes joined on to the part that looked like a turtle's back. The toes, however, were made up of little bony plates. And all the toes were close together, not free of one another as the fingers on her hand, or the toes on her foot. Why? Another question thought Mary—with no answer.

At any rate, she decided that the foot must belong to the crocodile.

The entire foot was about twelve inches long, and though

embedded in a thin slab of rock, it was still a heavy object. This time, however, Mary could not bear to leave such an exciting find to the whims of the weather. This treasure could not even wait for Joseph—she lugged it home herself.

And so her "crocodile corner" became richer by a few ribs and one foot.

Also, by this time, the sea had washed all the blue lias mud from the inside of the huge "honey and chocolate" ammonite. Now it was as beautiful as the first time they had seen it, the clusters of creamy crystals sparkling like diamonds within each curving chamber.

Mary reminded Joseph of his promise to find a way to reclaim it. With his mother's and Mary's permission, he used some of the money made from the sale of curiosities to hire a farmer's horse and cart one Sunday. Then with the help of some of his friends, he was able to bring it back.

That was a thrilling day, when Joseph and his friends, puffing and red-faced from the great weight of it, carried the enormous and showy ammonite into the shop.

"Joseph, how beautiful it is!" cried Mary, clapping her hands with delight. "It's like a dream come true. To have this beautiful curiosity where I can see it every day—and where it's safe from the tide. Put it on the table, in front of the window, so that it will be seen by everyone from outside."

Just then their mother came in. She stared at the spectacular ammonite in amazement.

"It really is as beautiful as you said it was, Mary."

This was high praise, coming from her mother. Although she had been pleased at the money brought in by Mary's curiosities, she did not usually find them interesting.

That night at church, Mary added to her prayers a special thanks for the magnificent "honey and chocolate" ammonite now reposing in the shop window, safe from the sea and the weather.

All next week there was hardly a villager—man, woman or child—who did not stop to admire and exclaim over the beautiful curiosity.

Mary was surprised to see a new respect on the faces of the children. She gave them a hesitant smile, and they asked politely if they could touch the lovely ammonite. Gladly she gave her permission. They began asking questions about the other curiosities. They listened to her answers with intense and respectful interest.

Joseph was right. The children were really interested in her "stones." She suddenly found it easy to be friendly.

It was a happy change for Mary.

X

MARY MEETS A SCIENTIST

ONE DAY NOT long afterwards, when Mary heard the London coach rumbling in, she rushed to the window to watch it as she usually did. She saw only one passenger, a gentleman, coming up the hill. He stopped to admire the large ammonite sparkling in the window. Then he looked closely at each and every item on the table. When Mary came to the door, he said, "I see that you are selling fossils."

"Fossils?" questioned Mary. "Oh no, sir. They are curiosities, sir."

The gentleman smiled, then explained. "Fossil is the correct word for curiosity." Then he added, "Haven't you better ones than these?" Mary looked at him in surprise. "Yes, sir. Inside. Please come in, sir."

She watched his amazement as he slowly looked around at the stony clutter inside the tiny shop.

"You certainly *do* have a supply—and a fine variety, too," he said with growing admiration. He began to inspect them one by one.

If only he knew what I have hidden away, thought Mary to herself.

This gentleman puzzled her. His interest in curiosities—no, he said they were fossils—was different from most of her customers. He inspected each one minutely. He even took a little glass from his pocket and looked again at them. From

time to time, his face lit up as though he had found something extraordinary. Finally he spoke.

"Would you call your father, little girl? I would like to buy a number of these specimens."

"My father is dead, sir."

"Who collects these fossils?" he asked in astonishment.

"I do, sir. And sometimes my brother Joseph."

"But you're only a young girl! How old are you?"

"I am eleven years old, sir—but I will soon be twelve."

"Incredible! Simply incredible!" and he shook his head in amazement.

"My father once collected curios—I mean fossils," Mary explained, then added sadly, "but he died several months ago. Now I am carrying on his work. On Sundays, my older brother, Joseph, helps me carry home the heavier curios—fossils. He also saws the honey and chocolate ammonites."

"Honey and chocolate ammonites! What do you mean?" the gentleman was dumbfounded.

"Like these, sir," and Mary pointed to several round the room.

"Oh! These ammonites have been replaced by calcite. Calcite is a mineral, a rather soft mineral. That is why your brother is able to saw through it.

"The metal ones would be too hard for him to cut. The metal, by the way is called pyrites, although many call it fool's gold."

Pyrites, calcite, I must remember these names, thought Mary to herself.

Then the gentleman walked over to observe the fish plaque that leaned against the wall. He studied it carefully for a time. "That is an unusual fossil fish. I have never seen one like it before."

When he realized that Mary was staring at him, he hastened to explain. "I am a scientist from the British Museum in

London. I am especially interested in the rocks that make up our earth, and in the fossils that occur in some rocks."

A sudden dawning lit Mary's face. "Oh sir, I remember my father said that there were men called scientists who studied the earth. He said they were trying to unlock its secrets."

"Nicely put, little girl. By the way, what is your name, may I ask?"

"Mary. Mary Anning," she offered shyly. Then forgetting her shyness, she exclaimed excitedly, "Sir, since you are a scientist, you will be able to answer all my questions. I have wondered so often about these curios—these fossils. How they turned to stone, and why, and—"

The scientist's face broke into a broad smile, as he held up his hand to interrupt her. "I can answer some of your questions, but not all. There are so many things we scientists haven't puzzled out yet.

"We think that the blue lias cliffs were once much lower, so low they were really at the bottom of the sea. The shales were mud then, and the limestones were just limey oozes.

"Creatures lived in the seas then, just as they do today. When they died, they sank to the bottom and were buried in the muds or limey oozes. Their soft parts decayed but the hard parts, such as bone or teeth or shell were preserved.

"Eventually, as the muds hardened into the rock shale, and the limey oozes became limestone, they entombed the remains of the once-living creatures. These remains we call fossils.

"In time, many of these remains were changed into stone, or a mineral—like calcite or pyrites. The bones in the blue lias, however, did not change. Sometimes the original shell on the ammonites can still be seen, too—like this one." The scientist picked up an ammonite that showed a thin layer of pearly shell in a few places.

Mary was annoyed that she had not noticed this herself. She *must* look at her finds more closely from now on.

"Then there were great earth movements. The rocks, with their fossils, were lifted from the bottom of the sea to become dry land. And that, we think, is the story of your blue lias cliffs."

Mary was completely astounded by all this information. "But how long ago did this happen, sir?"

"We don't know. We may never know. All we know is that it must have been a long time ago. Hundreds of years ago. Thousands of years ago. Maybe more."

There was silence again as the scientist returned to study the fossils and Mary thought over all that he had told her. Finally she spoke, "It's all so mysterious, isn't it, sir?"

"Yes, it is. Yet every time we find a new fossil, it helps to clear the mystery of the past a little more. We scientists are something like detectives, you know. Each fossil we find is a clue to the story of long ago. If we find enough clues, some day we may have the whole story.

"Now for instance, you have two new ammonites that I have never seen before."

"I have?" said Mary in amazement.

"Yes. You have two kinds of ammonites that are new to me. This one with little bumps on the ridges, and this flat one with no ridges showing at all." Mary had simply taken these differences for granted.

The scientist went on. "And have you noticed the beautiful wavy line that is between the ridges of many of the ammonites, and also the metal ones?"

Mary shook her head, half-ashamed to admit again, that there was something she had not seen on her beloved treasures. Sure enough, a faint wavy leaf-like design was repeated again and again on the surface. It was even lovelier seen under the scientist's glass, for the glass magnified everything.

"We think that each species of ammonite had a different pattern," added the scientist.

This seemed a good time to ask the scientist another question. "Why are the ammonites divided into so many chambers, sir?"

"That's easy, child. The soft-bodied animal first lived in the smallest chamber. Then as it grew larger, it built a larger

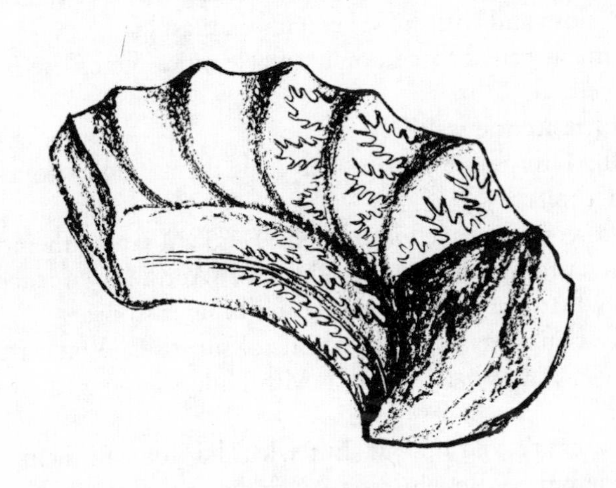

chamber. It did this many times as it grew, building a roomier chamber each time and leaving the others empty."

"Do ammonites live anywhere in the world today?" asked Mary.

"No, they have become extinct. Why, we don't know. The nearest relative living today is the chambered nautilus.

Its shell is of mother-of-pearl and it builds chambers the same as the ammonite did. But the chambered nautilus lives only in warm seas.

"Some of the other fossilized creatures you have found are still living—starfish, sea urchins, snails, cockles."

"Yes, I've noticed that, sir," admitted Mary.

"Now, why they should still live and not the ammonites—well, it's just another mystery that we scientists may never be able to solve."

Then the scientist went on with his inspection of the fossils. Every now and then he set certain ones aside.

"I am surprised that I don't see belemnites here," he finally observed.

"What are they, sir?"

"Oh, I forgot. You probably call them finger stones, or thunderbolts."

"Oh, yes sir. I have lots of them. But I am saving them for a special customer, Miss Annie Philpots." And Mary opened the drawer that was full to bursting with finger stones.

The scientist gave a little whistle of surprise. "You certainly do have a lot. What on earth is Miss Philpots going to do with them?"

"She hasn't told me yet. But what did you call them, sir?"

"Belemnites. Bel-em-nites." Another new word she must remember.

"Do you know what they were, sir?"

"Yes. They were the inside skeleton of an animal that looked something like a squid. But, like the ammonites, belemnites no longer live.

"I haven't seen any crinoids in your collection, either," commented the scientist. "Come to think of it, though, you probably call them sea-lilies."

"Sea-lilies? In this drawer, sir."

Again the visitor whistled in surprise when Mary opened

the drawer. "You have some very fine ones here. In fact, here is a sea-lily that is new to me."

"The sea-lily must have been a beautiful flower when it lived," observed Mary.

"Oh no, child. The sea-lily was a *creature*."

As Mary looked at him in bewilderment, he hastened to explain further.

"Now mind you, although it was a creature, it did look like a flower. Its stem had a root that fastened it to the bottom of the sea, or perhaps to a rock, or even to a bit of seaweed. On the other end of the stem was the head with its many long arms. These arms waved backwards and forwards, and washed food into the mouth. The mouth was on the top of the head. Sometimes this mouth can be seen in fossil crinoids."

So the proper name for a sea-lily is crinoid, thought Mary, but I like the name sea-lily much better. It's a prettier name.

"Does the—the crinoid still live today?" She decided it was better to use this name in front of the scientist.

"Yes, but only where the sea is shallow and warm. That is why we think the seas here must have once been warmer than they are now.

"By the way, the sea-lily was related to the starfish, you know."

"Oh?" That was a surprise to Mary.

The scientist paused, then asked, "Tell me, Miss Mary, do you have any more treasures hidden away?"

"Oh sir, that's what I call them too," said Mary, pleased that this great scientist would call curios—no, fossils—by her favourite name. She showed him her most secret place, the crocodile corner. When he looked in, he gave a gasp of astonishment.

"Do you really think these belonged to a crocodile, sir?"

"We don't know. All I have ever seen are the vertebrae. But you've found so much more. According to the largest vertebra here, the animal must have been much larger than we had thought. But the most curious thing is this skeleton of a foot, or whatever it is."

He studied it for a long while in silence. "Nothing like this has ever been seen before. It can't be a foot. The toe bones are too close together. Perhaps it was covered with skin like a flipper." Then he was silent again as he studied it further.

"Mary," he finally said, "you seem to have parts of an entirely new animal here. You must keep a close watch for more of it. If you ever find enough of the skeleton for us to identify, it will be an exciting day for you, for me, and for my fellow scientists.

"You see, Mary, you can be a wonderful help to us scientists. We haven't time to go collecting as often as you do and naturally you will find more than we do."

Mary could scarcely believe that she, a young girl, could be of real help to important people like scientists.

"And if you do find the skeleton, will you let me know at once?" he asked. He wrote his name and address on a slip of paper. "You are able to read and write, I hope?" he added anxiously.

"Oh yes, sir." How thankful Mary was that she could say that. After all, most poor children like herself hadn't the slightest idea how to do either. "The only thing is," said Mary, "if I find the whole skeleton, I have already promised it to Mr. Henley. He is the Lord of the Manor and he is a buyer of fossils for a small museum in London."

Disappointment clouded the scientist's face for a moment, then he brightened up. "I'll speak to him and see if we can come to some arrangement."

"If this animal is not a crocodile, what shall I call it?" asked Mary.

"Better just keep on calling it a crocodile until we know more about it."

The scientist asked her always to set aside for him anything unusual, especially bones. He repeated again how important she could be to him and his co-workers in their efforts to learn about the past.

Then he paid for the many fossils he had chosen. It was the largest sale Mary had ever made to one person. And to think all these fossils she had collected would become a part of the British Museum, the largest museum in England!

XI

A NEW FRIEND

MARY WENT ABOUT on a cloud for days after the scientist's visit. She could scarcely believe that she—a poor, young girl, with very little schooling—could help play an important part in helping scientists solve some of the mysteries of the past. It didn't seem possible!

Neither did it seem possible to her mother that her young daughter and her "stones" should be of such importance to someone so learned.

To Joseph, it was no surprise at all. "And when you find that crocodile, or whatever it is, just wait and see what excitement there will be *then.*"

"*If* I ever find it," said Mary.

Soon after the scientist's visit, Mr. Henley called one day to congratulate her. "So you are a scientist's helper now. Splendid!" Then he added with a smile, "If you ever find that crocodile, it's going to be a great day for our town."

"But if I find it, the scientist said that he would like to have it, too." Mary's forehead furrowed with concern.

"Oh, we've come to an understanding about that, Mary. He and his fellow scientists will study it for as long as they wish. Then it will be mine. You must keep looking for it. It must be there somewhere in those blue lias cliffs."

Mr. Henley smiled at her, then tipped his hat in farewell and set off up the hill. As Mary watched him, she had a nice, warm feeling creep over her. Mr. Henley, the most important man in the town, was her friend.

The holiday-makers soon would be flocking into the town and Mary did not have much longer to collect. The stormy season was well past and the mud-slide yielded few fossils. Because of this, she and Tray explored far and wide along the coast.

Always she looked for the "crocodile." Although she found some odds and ends of it, they were identical to parts she already had.

Here and there along the top of the hills were quarries where men dug out the limestone. She even asked permission to look for fossils in these quarries. She also asked the quarrymen to be on the lookout for any special fossils. She hesitated to tell them that she was hoping to find the skeleton of some strange animal. They might just laugh at her.

During this time, she found her most handsome ammonite yet. It was as large round as the bottom of their tea kettle. It was of grey limestone, with curving ridges that stood out sharply and smoothly. In between the ridges, the beautiful leaf-like pattern showed up clearly and distinctly.

Mary did not want to part with this precious find, not for one moment, so she carried it home, herself. Many times she stopped to rest her aching back and arm muscles, while Tray licked her face in sympathy. But at last the exhausting journey was over and the ammonite safely in the shop. Then and there, Mary made a firm resolution—she would never sell this stone, not even to the scientist, for she might never find another as beautiful nor as perfect. She tucked it in a far corner where it could not be seen by prying eyes.

Something else wonderful happened to her about this time.

One day as she searched along the beach, her eyes always cast downward, she bumped head on into a young boy. Both rubbed their throbbing skulls ruefully, then both broke into laughter.

"Were you looking for fossils, too?" asked the boy with a grin, still rubbing his head. He appeared to be about Joseph's age but was well-dressed, like a gentleman's son.

"Yes," answered Mary, suddenly shy and tongue-tied.

"What is your name?" he asked.

"Mary Anning."

"I am Henry De la Beche. My mother and I have just moved to Lyme Regis and I have become interested in collecting fossils. The only thing is, I get so interested in looking for them, I never see the sky or the hills or any of the scenery around me—or people," and he rubbed his head again, with a wider grin.

Mary had not thought of it like that before. Come to think of it, rarely did she look above and beyond the rocks of the cliff or the stones on the beach. No one else would have understood this, except someone as keenly interested in hunting fossils as herself. How glad she was she knew that word, fossil. She would have sounded very ignorant using the villager's word, curiosity.

"I have never heard of a girl collecting fossils before," he commented.

"I learned from my father. He—he died last autumn," explained Mary.

"I'm so sorry," he said quickly in sympathy. "And what do you do with your fossils?"

"I sell most of them."

"You do!" He seemed amazed.

"Yes, I sell them from my father's workshop."

Suddenly Mary wondered about her appearance. She hoped that her face was not streaked with blue Lias mud. She felt her sun-bonnet hanging down her back. Quickly she slipped it back on her head and retied the strings.

Then the two began searching again, but going in the same direction this time. It was not long before Henry De la Beche spoke up admiringly, "Your eyes are certainly sharper than mine, Mary, for you find so much more than I."

"That's because I've collected much longer than you," explained Mary. "Wait until you've searched the beach as often as I have!"

At that moment, she spotted a group of tiny fossil fish within a small area of the rocks. Quickly she dropped to her knees and, slowly and carefully, began working the fish out with her

chisels and hammer. So intent was she on her work, she almost forgot she was not alone until Henry spoke up.

"How skilful you are with your tools."

"Oh, thank you," answered Mary. To be praised and respected by this young stranger gave her a new and very pleasant feeling.

Not so, Tray, however. He began to growl and show his teeth to the newcomer.

"What's wrong with him?" asked Henry in surprise.

Mary just blushed pink and dropped her eyes.

Henry stared at the little dog quizzically. Then his face lit up. "Oh, oh. Now I know. He's jealous of me."

"I'm afraid so," admitted Mary. "You see, Tray has been my only companion on so many of my collecting expeditions."

"Come on, Tray," pleaded Henry. "Let's be friends."

After Henry had patted and fussed over Tray a little, Mary was relieved to see him stop growling and snuggle up to Henry. His jealousy was over.

From then on, Mary and Henry De la Beche often met on their collecting expeditions. During the course of their conversations, she learned that he hoped to become an officer in the army; and in preparation for this, he would soon be attending the Royal Military College nearby.

Knowing that Mary was helping to support her family by selling fossils, he generously gave her many of the ones he himself found. For the first time in her life, Mary had a real friend near her own age, one who shared her interests. Gone now was the bleak loneliness left by her father's death.

XII

SELLING FOSSILS

THE SUMMER SEASON was close at hand and Mary soon would have to stay in the shop every day.

To her mother's and Joseph's amusement, she kept changing her sign on the table outside. She would like to have changed the word, curiosities, to the more proper word, fossils. But she felt that would be too strange a word for most of her customers.

Underneath CURIOSITIES FOR SALE she added FOR SOUVENIRS AND ORNAMENTS. When she noticed that more and more people bought the dainty metal fossils to set in brooches, she added AND JEWELLERY.

One day when the door of the shop persisted in swinging shut, she finally placed a large ammonite against it as a door-stop. Her next customer was so taken with the idea that he bought one for the same purpose. And so, AND DOOR-STOPS, was added on the sign.

When a gentleman in the village bought her largest ammonite to decorate his rock garden and another used one to place in his stone wall, she started all over again and made a new sign.

CURIOSITIES FOR SALE
for

SOUVENIRS — ROCK GARDENS
JEWELLERY — DOORSTOPS
ORNAMENTS — WALLS

The wind whipping in the open doorway almost blew her sign away as she worked on it. Without thinking, she picked up the nearest fossil, a sea urchin, and placed it on the far corner of the sign, to anchor it. Later, when she realized what she had done, she added to the bottom of the sign AND PAPER WEIGHTS.

Finally, the first day of summer arrived.

As Mary stood ready and waiting for business, she looked round at the workshop fairly bursting with fossils of all kinds, of all shapes and sizes. How lucky her customers would be. All they had to do was choose the ones they liked and pay for them, all in the space of a few minutes.

At that point in her thoughts, in walked her first customer of the season—that is, she hoped that she was a customer. She was a stout red-faced woman, well-dressed. She must be a holiday-maker, perhaps come for the sea bathing for which Lyme Regis was so famous. Mary stepped forward with a smile.

But there was no answering smile. The woman merely grunted. She spent a long time poking among the neatly arranged fossils, until she had completely disarranged them.

"Your prices are too high—far too high," she complained in a loud, disagreeable voice. "After all, you only have to gather these stones off the beach."

Mary's mouth fell open. Did she really think it was as easy as *that*? She thought of the many miles she had trudged on the beach to collect them, in all kinds of weather; of the heavy loads she had carried and how weary she had often been, her arms aching and her feet sore.

She remembered the tedious hours spent washing and scrubbing the fossils; the hard work of sawing and polishing the "honey and chocolate" ammonites, and the polishing of the metal ones.

Suddenly Mary had an inspiration. "Ma'am, you might prefer to hunt for your own curiosities. After all, the cliffs are full of them. Then they would not cost you anything."

"Humph!" growled the woman ungraciously, "I think that is what I shall do. Where should I go?"

"Oh, anywhere along the coast," said Mary wide-eyed, trying to look innocent.

Away went the woman with high hopes. She set off to the west, as Mary had hoped she would. There were plenty of huge ammonites there, goodness knows, but they were all far too large to be carried away.

The woman apparently did not realize the tide was coming in, nor had she appeared to notice the dark clouds drifting near the village.

Also she had no tools.

Standing in the doorway, Mary watched her until she had disappeared from sight. Then she set to work to restore order to her jumbled up fossils.

Just as she expected, the woman returned some time later, a much changed person, very tired and rather frightened. She was wet and bedraggled from the quick summer shower that had swept along the coast. Her shoes were badly scuffed. She had obviously fallen, for the skirt of her dress was torn and her forearm bore angry red streaks.

And—she was empty handed.

There was no loud complaining this time. She quickly chose a few of the best fossils, paid for them without a word and left.

That night at supper, Joseph and her mother listened intently as Mary told how she had dealt with her first customer of the summer.

"Selling fossils isn't going to be much fun, if all the visitors are like that," she said with a sigh.

"Cheer up, Mary," consoled Joseph. "You'll probably never meet another person like her."

Joseph was right. That summer sped by swiftly, pleasantly and profitably.

One morning Miss Annie Philpots stopped at the shop and said in a mysterious tone, "Mary, could you come for tea this afternoon? I have something to show you."

To be invited to a gentlewoman's house was a great honour. Mary was thrilled. When her mother heard of the invitation. she offered to take her place in the shop while she was away.

At last it was time to go. A curious and excited Mary made her way up the steep main street to the cottage where the Philpots ladies lived. It was a pretty cottage, topped with a thatched roof and surrounded by a neat and colourful garden.

Miss Annie answered her knock. Her hair was piled high on her head except for a few ringlets that prettily framed her face. She wore a delicate pink silk dress with a graceful lacy shawl collar, that was fastened with—here Mary gave a start—the brassy fragment of sea-lily stems she had bought from Mary. How lovely it looked.

Miss Annie welcomed her warmly, then led her into a large sitting room. Never before had Mary been in such a colourful room!

Gay flowered linen decorated the upholstered chairs and couches and hung at the windows. A red rug, thick and luxurious, covered the floor. The fireplace was of elaborately carved marble.

Miss Annie motioned Mary to a chair. Mary was almost afraid to sit on the lush upholstery, so she perched shyly on the edge of the chair.

"And what do you think of my picture?" asked Miss Annie eagerly.

Mary was so fascinated by the beauty and luxury of the whole room, that she was puzzled at first by Miss Annie's question. Then she saw that Miss Annie was looking at a large framed picture above the fireplace. No, it wasn't really a picture. It was a beautiful and intricate design of graceful and curving lines. How odd.

Mary peered at the picture more closely. Why the design was made up of scores of finger stones—large ones, medium-size ones and little ones.

"Oh, ma'am, so that's what you were doing with all the finger stones!" Mary had forgotten her shyness. "I've wondered so often."

Mary continued to study the picture. To think that a design of such grace and beauty could be carried out with the smooth pointed finger stones. How glad she was that she had persisted in collecting them.

"They make a lovely picture, Miss Annie—so much nicer than being heaped up in a drawer—or left on the beach for the tide to carry out."

"It's all thanks to you, Mary. You must have walked many miles collecting all those finger stones," said Miss Annie gratefully.

Then Mary saw sitting on a small side table, a rock covered with clusters of small "honey and chocolate" ammonites. She remembered selling it to Miss Annie.

"I like to use the most beautiful specimens for ornaments," explained Miss Annie.

As Mary looked round the room, she saw more tucked in with the fine china and glass—a sea urchin on another table, a few snails on a corner shelf, a starfish in a wall cabinet. How much better they looked here than in the stony clutter of her shop.

Then Miss Annie led her into a small side room to show her the tall bureau of many drawers in which she kept most of her fossils. She opened the drawers one at a time. Altogether there were hundreds of fossils neatly stored away.

Although most had been bought from her father in years gone by, Mary recognized with delight many that she herself had collected. It was most satisfying to know that the fossils she and her father had collected were so cherished and so carefully stored in this beautiful house.

Miss Annie's sister, Miss Margaret, now entered the room with a heavily laden silver tea tray. "I think it's time you two eager collectors had a cup of tea," she announced gaily, leading the way back to the sitting-room.

Mary felt like a queen, sitting in a soft chair, being given tea in the finest of china cups, tea that had been poured from a handsome silver teapot. The Misses Philpots behaved as if she were one of them, instead of the daughter of a poor carpenter and his wife.

Mary left for home in a glow.

Her mother met her at the door in great excitement. "Mr. Gloster was here, Mary. He wanted the best and the largest ammonite you had. We searched all through the shop and finally he found the perfect one, in that corner. He was so pleased!"

Mary's heart sank. They must have found her most prized ammonite, the one that she never intended to part with. She looked in the corner. Yes, it was gone. She would never find another like it. Tears came to her eyes.

"Mary, whatever is the matter? I thought you would be pleased."

"I never intended to sell that ammonite, Mother," wailed Mary. "It was so perfect. I'll never find another like it."

"Would it help, if you knew that Mr. Gloster was going to place it in the arch above the main door of his lovely new house?" asked her mother.

Mr. Gloster was a rich gentleman who was building a grand new house overlooking the promenade by the sea. As Mary often went walking on the promenade, she realized she would be able to see her treasured ammonite whenever she wished. Come to think of it, many other people could enjoy it.

"Oh, that's different." Mary gave a sigh of relief. "I was afraid I would never see it again."

Although many strangers bought her fossils and took them away to distant places, it suddenly occurred to her that a considerable number of villagers had bought some of her finest ones. In fact, she would find at least one rock garden, garden wall or doorstep decorated with her fossils in whatever street she now walked down; while inside many houses, she knew that they sat on top of fireplace mantels and window sills. It was a very nice feeling to know that so many remained within the town.

In the meantime, her shop was beginning to look like Mother Hubbard's cupboard, for it had been a good summer. So good, in fact, that the Anning family often had meat twice a week, something they had never been able to afford before.

XIII

AN EXCITING DISCOVERY

It was autumn again. Mary was glad to be walking along the lonely beach once more, searching for her fossil treasures. Even Tray seemed to be happier. Until the storms came, of course, she could not expect to find too much. She doubted if anyone had ever hoped for bad weather more than she.

Late in November her hopes were fulfilled when the stormy season set in; but it also meant that for days on end, she and Tray could not go collecting. The tides were much higher during the autumn gales, as the winds whipped the waves into a fury and sent them crashing high on the beach.

The stronger the winds blew, the harder the rains beat down and the more fiercely the waves pounded the beach, the gloomier became the villagers. They said that they had never remembered a worse storm. It would do terrible damage along the coast.

Mary was delighted. Many fossils were sure to turn up now.

When the storms finally subsided and it was safe to go on the beach, she and Tray set out. As she had expected, there were many changes. A projecting ledge of Church Cliff had crashed down. She would have dearly loved to search through the jumble of rocks spread out below but she knew that it would be far too dangerous. Water still trickled down from the cliffs and at any moment could trigger off another rock-fall.

When she reached Black Ven, she found that the storm had

wrought even greater damage to its steep front. The beach below was littered far and wide with broken rock. It, too, would have been another good hunting ground.

Ahead she could see that the mud-slide stood much higher than last year and that it had spread much wider. Even before she reached it, she was busily picking up fossils scattered everywhere on the beach. In fact, her basket was nearly half full by the time she started looking on the slide itself. It was a thrilling time to be collecting.

Even Tray seemed to share her joy. Each time she exclaimed aloud over some special find, he echoed her pleasure with happy, staccato barks. No longer was he a nuisance, but an enjoyable companion.

Never before had Mary's basket filled up so fast; nor had her stock-pile of fossils for Joseph to carry been so large. Of her "crocodile" alone, she found several vertebrae and another flipper, as the scientist had called the foot of the mysterious animal. Would she ever find enough of its bones for the scientist to identify?

As the days went by, her shop no longer resembled Mother Hubbard's cupboard. It was near to overflowing. Neither was there any lack of sawing for Joseph or polishing for her, as she brought home a wealth of "chocolate and honey" ammonites. Although she knew their proper name was calcite ammonites, she was loath to give up her own pet name for them.

Mary now realized she was finding many varieties of ammonites. Some were flat. Some were bulging. Some were ridged and others were smooth. Some were beautifully ornamented with leaf-like designs, while others had no trace of decoration. Similarly with the other fossils. All came in many varieties. Whenever she found something new or different, she set it aside for the scientist.

The "crocodile" remains bulged out from their corner. In addition to many more vertebrae, there were more ribs and two more flippers—and most exciting of all—part of a long narrow jaw filled with many conical-shaped teeth.

Mary saw Henry De la Beche only on Sundays for he was attending the Royal Military College. Even though he was a gentleman's son and could have been riding or hunting with his friends, he much preferred joining Mary and Joseph in their searches along the sea coast. When the three of them went

together, much more was found; and as usual, Henry insisted on giving many of the fossils he found to Mary.

One Sunday when the fossils were scarce on the mud-slide, they turned homewards much sooner than usual. As they passed by the clutter of stones at the foot of Black Ven, Mary said wistfully, "I wish it was safe to go there."

Joseph stopped short and considered the cliffs carefully. "There's no water running over the cliff. Everything that's loose at the top must surely have tumbled over by now. I wonder—if one of us stood guard, keeping an eye on the top of the cliff, why couldn't the other two search?"

"That's an idea," agreed Henry. "Let's do that."

Henry offered to stand guard first. Joseph and Mary eagerly began investigating. As Mary had suspected there were many fossil treasures. Every Sunday, from then on, if there had been no recent storms, they searched below both Black Ven and Church Cliff, with always one of them on guard.

So experienced had Henry become at finding and prying out fossils that one day he exclaimed, "You know, I am beginning to wonder if I should become a scientist instead of a soldier?"

"Oh, Henry, why don't you?" cried Mary excitedly.

"Good idea, Henry. Then instead of dodging bullets, you could dodge rock-falls," grinned Joseph.

"Joseph!" scolded Mary.

One Sunday, when Mary was ill with a severe cold and Henry was away with his parents, Joseph went hunting alone with Tray. It was nearly dark when he and Tray returned home. Although he was covered with blue lias dust and his feet were soaking wet, he was grinning triumphantly.

Mary sat huddled on a stool close to the crackling fire for warmth while her mother was getting supper ready. Seeing the unusual look on Joseph's face, she cried out, "Joseph! You've found something special. Let me see."

Without a word, he pulled from under his jacket a large, flat skull. It was nearly as long as his arm. The eye sockets were enormous. The long, narrow jaw was full of cone-shaped teeth.

"Joseph! The crocodile head!"

Joseph smiled proudly.

"How did—where—oh, tell me everything!"

Joseph laughed out loud at his sister's impatience. Their mother had stopped her work and was staring at the skull in utter amazement. Then she and Mary sat down on the other side of the fireplace to hear his story.

"Well, I wasn't having much luck," Joseph explained, "so I decided to give up and return home. I was passing Church Cliff when I saw something strange in the face of the cliff. As I went over to look at it, guess who stood guard, without saying a word?"

"Tray, you darling. It was you." Mary reached down and pulled the little dog close to her.

"The strange thing I saw was a jaw," Joseph went on. "As I pushed away loose earth and bits of rock, I could see there was a whole skull. It was further back and tightly wedged between two layers of rock. I knew I would need help to get it out. There was also about two hours before the tide would be in.

"I told Tray to stay on guard so I wouldn't lose time finding the right spot when I returned.

"I raced up the path on the side of Church Cliff and then on up the hill to the quarryman's cottage. I asked him if he and his sons would come with crowbars and help me pry out the skull. Then we all ran back. I don't think Tray had moved an inch. We were soon at work, prying and tugging. The tide was coming in. Before we got the skull out, we were standing ankle deep in water. It was a good thing we were close to the town.

"And Mary—when we pulled out the skull, it broke off from something else. Probably the backbone. I think the remainder of your crocodile may be there."

"Oh Joseph, I do hope you are right." Mary's eyes sparkled at the very thought. "I can hardly wait to go looking for it."

Mrs. Anning spoke up quickly, "But, Mary, I don't think it will be safe for you to work under the cliff."

"Mother's right, Mary," added Joseph. "Besides, you will

need help to get it out. You'll have to go to the quarrymen and ask them to help you as I did."

It was an impatient Mary that went to bed that night.

By midnight, however, a terrible storm sprang up. Lightning stabbed the darkness with blinding flashes. Ear-splitting thunder rolled again and again across the heavens. Then the rains began. All night they poured down in torrents. And the next day. And the next night. And the day following. Mary was getting more impatient every hour.

In the meantime, word had spread that Joseph had found a crocodile skull. Nearly every villager found time to drop in and marvel at it.

"I've heard my grandfather tell about a skull like that, but I've never laid eyes on such a thing before," said one.

The others nodded their heads in agreement. They, too, had often heard of such skulls but had never seen them.

"Is it really a *crocodile*'s head?" asked another.

"Well, that's what it was always called," someone else said.

This explanation seemed to satisfy everyone, that is, everyone except Mary. She would wait until the scientist had passed judgment on it.

At last the weather calmed down and Mary and Tray hurried off to Church Cliff. What did she see but tons of rock covering the lower part of the sloping cliff and the beach below. It was an enormous rock-fall, triggered off by the heavy rains. If her "crocodile" was there, it was well hidden now. There was nothing to do but wait until another storm came and carried away the rubble.

Mary was bitterly disappointed.

She and Tray moved on until they reached the mud-slide. If the storm had been her enemy at Church Cliff it had been a generous friend here. The collecting was so good that she almost forgot her buried "crocodile."

The collecting continued to be good for weeks afterwards,

until the shop was cluttered once more with a wonderful stock of fossils. Joseph had built her more shelves on the walls. On the topmost one, perched her most prized possession, the "crocodile" skull. It was still the wonder of the village and of strangers who dropped in. Would she ever find the remainder of it? That question was constantly in her mind.

XIV

SKELETON IN THE ROCKS

AT LONG LAST, another storm—a great gale this time—blew up. Day after day, for a week, the wind blew hard from the southwest. It sent the waves crashing much higher than usual on the beach.

Surely all the loose rock at Church Cliff will be carried away now, Mary said to herself.

As soon as it was possible to go along the beach, she and Tray set off. It was early in the morning, just before the village came to life. To her great joy, all the clutter of rock at Church Cliff had been swept out to sea and no new rock-falls had occurred. The slopes were as clean as a pin and the top of the ledge appeared solid and safe.

Without even going close, Mary could see part of a backbone showing. The storm and the tide had certainly been her friend this time. She commanded Tray to stand guard directly opposite the backbone. Then she ran as fast as she could up the cliff-side path and on up through the town to the quarryman's cottage. He and his sons were just getting ready to set out for work.

"Oh, sir," Mary was puffing so hard that she could scarcely get her words out, "I have found—the crocodile's backbone—on Church Cliff. But it is too large—for me to get it out by myself. Could I pay you and your sons—to take it out for me?"

The quarryman nodded his assent.

"You will need more men, I think," added Mary.

The quarryman quickly asked one of his sons to go for more help. The other he sent to the top of Church Cliff to look for loose rock before they began working below. Then he gathered up all the hammers, crowbars and picks that he could find and set off with Mary down through the town and back to the base of Church Cliff.

Loyal little Tray was still on guard exactly where Mary had left him. Before long, the quarryman, his sons and the other helpers were hard at work.

"Please be careful with the bones," Mary called out to them, "they're brittle and break easily."

The men said nothing but nodded in agreement. They worked quickly and carefully, as carefully as Mary would have done herself.

Word soon spread through the town that Mary had found the crocodile skeleton. Everyone who could, came to watch—the Misses Philpots, shopkeepers, farmers, housewives, chimney sweeps, the candlestick maker and even the muffin man. Mr. Henley, unfortunately, was away.

Little by little, the quarrymen removed the rock layer above where the backbone lay. Suddenly one of them shouted, "The whole backbone is here!"

A few minutes later a second quarryman shouted, "I can see the ribs—many ribs!"

It was not long before a third cried out, "I see two feet! The front feet!" Then came an astonished cry, "The two hind feet are here, too—but they are much *smaller*!"

The eager crowd on the beach could barely contain their curiosity but it was some time before this curiosity could be satisfied. The rock in which the skeleton was embedded was so large that it could not be taken out in one piece. It had to be removed in several pieces. As each was taken out, it was brought down and placed before Mary.

Eager hands helped her fit them together again. Finally the entire skeleton lay at her feet. It was much longer than she was tall. Everybody was silent for a time as they studied it.

As one of the quarrymen had first observed, the front flippers were much larger than the hind ones. Also, at the end of the backbone, the impression of a shark-like tail fin could be seen in the rock. An animal with both flippers and a tail fin?

"What a strange crocodile!" muttered the villagers.

"It's head is like a crocodile's—but certainly the body isn't," said someone.

Mary said nothing. It was clear to her now, as it had been to the scientist before, that it could not possibly be a crocodile. She must send word to the scientist today.

Then everybody tried to help carry the skeleton. A joyful procession walked back to the town, Mary and Tray leading the way. It was a happy day for the people of Lyme Regis. The crocodile that they had heard of for so long, but had never seen, was now found. Everybody talked at once. Their town would be famous. Mary would be famous.

Someone suddenly had a wonderful inspiration. It was like the story of "Beauty and the Beast." The crocodile was the Beast and Mary was the Beauty.

Mary blushed deeply.

"But I didn't find the crocodile really," she protested vigorously. "Joseph found the head and he thought the skeleton might be there, too."

The villagers refused to listen to her. They cheered and teased her fondly by turns.

As the merry procession neared the shop, Mary saw her mother appear at the doorway in surprise.

"Why Mary," she exclaimed, "I knew you had found the skeleton, but I had no idea that so many people would be interested in it."

The slabs of rock containing the skeleton were carried into

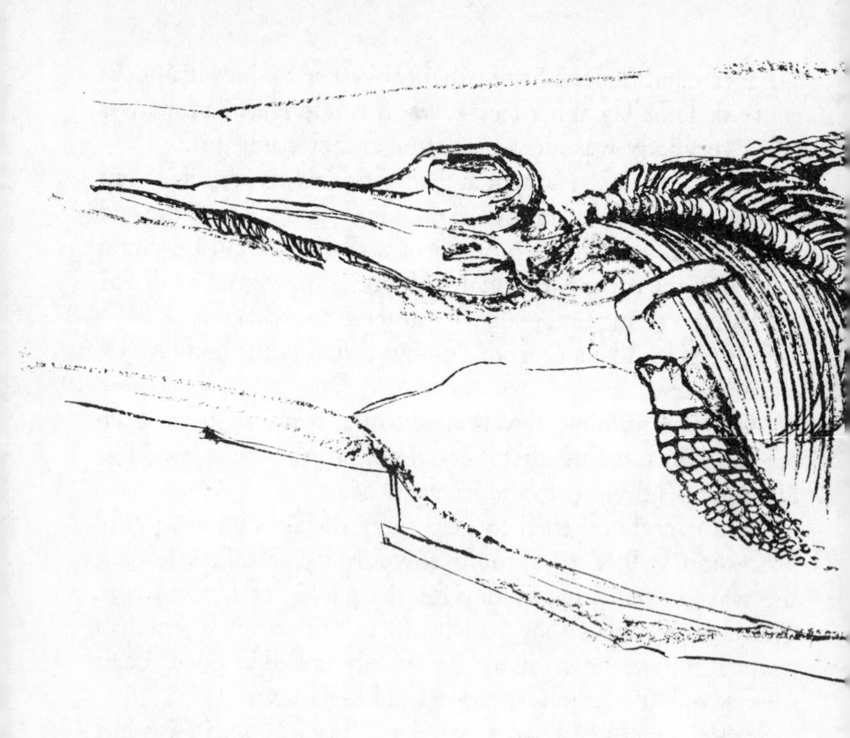

the shop and placed on the floor. Then they were fitted together, like a jigsaw puzzle.

Someone reached up and added the head. So large was the whole skeleton that there was barely enough floor-space for it. Then someone measured it, from the tip of the snout to the end of the tail. It was ten feet long! Over twice as tall as Mary.

For the remainder of the day people streamed in and out to look at the bones of the new-found monster that had lived long, long ago. Almost the last person to see it that day was Joseph. After he had admired and marvelled at it, Mary told him everything that had gone on. Then with great concern, she added, "But Joseph, people keep saying I found it when it was really you."

"Not really, Mary. You started the whole thing. If it hadn't been for you setting up a business in fossils, I wouldn't have been looking for it. Remember the day on the beach when I predicted you'd be famous for collecting fossils?"

"But you were only teasing, Joseph."

"Yes, but you wait until your scientist and his friends see this skeleton. You'll be famous all right."

Mary was not convinced. "I don't think anybody becomes famous at *my* age," she said doubtfully.

Before she went to bed that night, Mary wrote a letter to the scientist.

XV

MARY IS FAMOUS

JOSEPH WAS RIGHT again. As word of the monster spread beyond the town and to the newspapers in London, Mary became famous throughout the land.

Mary knew that Mr. Henley, who was visiting in the next county, would probably read the exciting news in the newspapers, so she was not surprised two days later to see him jump off the London coach, almost before it had stopped. As he hurried up the hill to the shop, she stood in the doorway smiling.

"Well, Mary, I knew you'd find it!" he exclaimed. Then he looked at the immense skeleton stretched across the shop floor. "What a size it is!"

Thoughtfully he studied it for some time. Finally he spoke. "No wonder the scientists didn't think it was a crocodile—with those feet and that fish-like tail fin. But I wonder what it is?"

"I sent word to the scientist," said Mary, "and he will be coming soon to see it."

"And then the monster will be mine?" he smiled at her questioningly.

"Of course, sir."

The scientist from the British Museum in London, with his team of assistants, descended upon Lyme Regis and Mary's shop. They congratulated her warmly on her strange find. Then they set to work to examine it carefully, bone by bone.

They also made an exact drawing of it, so that they could study it when they returned to the museum.

"This was not a crocodile, as I suspected, Mary," said the scientist when they had finished. "It appears to have been a great fish-like reptile. Because of the tail fin, it would have been able to swim like a fish. We think the feet, or fins perhaps, were covered with skin like flippers. They would have helped to balance the animal. Unlike a fish, however, it had lungs and breathed air.

"It is a very exciting discovery that you have made, Mary, for we never before knew such an animal existed.

"A brand new name will have to be given to it. When we decide what this name is, I'll let you know. Then we will send word to other scientists round the world of this amazing discovery by a twelve-year-old girl, Mary Anning of Lyme Regis, in the year 1811."

Mary was terribly proud of such a wonderful honour but she felt very shy in front of these learned gentlemen. She did not know what to say.

"If this strange animal lived long ago," continued the scientist, "there are probably others. We hope that you will keep on the look-out for them."

This was easy to answer and Mary spoke up quickly, "I will, sir, I will."

"And from now on, Mary, we would like to have the first chance at any new monster you may find," said the scientist smiling.

Mary assured him that she would let him know first, if it happened again.

Then the scientist went over the pile of special fossils that she had set aside. He found them all of extreme interest. Before he and his assistants departed, he had bought the whole lot.

That night Mr. Henley came to claim his new-found reptile. He paid Mary the staggering sum of twenty-three pounds for it. She was astonished. Why, it would have taken Joseph many weeks to earn so much money. How pleased her mother and Joseph would be.

There was great rejoicing at supper that night.

"You know, Mary," Joseph announced, "it looks as if I should be able to leave the upholstery shop soon and become your full-time partner."

"Oh, Joseph, how wonderful that would be!" Mary clapped her hands with joy.

"I never dreamed that collecting and selling fossils could be a full-time business," said Mrs. Anning, shaking her head in wonder.

Later the scientist informed Mary that the name, Ichthyosaurus, had been given to the new fossil reptile. It was a tongue-twisting name but he explained that it was a Greek word meaning "fish-like reptile." He even told her how to pronounce it—ik-thee-o-sore-us.

He said that the ichthyosaurus was now on view at the small London museum to which Mr. Henley had sold it; and that it was attracting attention from far and wide.

To Mary's surprise, people even came to Lyme Regis to see *her*, the young girl who had made the famous discovery. It was pleasant but embarrassing.

More interesting to her were the famous scientists of the day who dropped in to look at and buy from her stock of fossils and who often went out collecting with her. Then there were other well-known scientists from far away who wrote and asked her to provide them with specific fossils.

One night after a letter had come from a scientist in Switzerland, Joseph commented gleefully, "You're really famous now, Mary girl, when people write to you from as far away as Switzerland."

"Mary girl." Joseph's use of her father's pet name startled her. She had not heard it since he'd died. How proud of her he would have been; and how grateful she was that he had introduced her to this exciting work.

And what of the future? Mary sighed hopefully, "Oh, Joseph, I wonder how many more wonderful fossils are hidden in the blue lias. You know that largest vertebra I have? It's much larger than any of the vertebrae in the ichthyosaurus. There must be a larger animal yet to be found."

She stared dreamily into space for a few moments, then she added excitedly, "And, Joseph, maybe I'll find my lovely golden snail again some day."

Suddenly Tray began barking and jumped up and down in front of her to catch her attention.

"Oh forgive me, dear Tray," Mary said as she patted him fondly, "Maybe *you* and I will find the golden snail."

EPILOGUE

WITH JOSEPH'S ASSISTANCE, Mary continued to hunt for fossils and support her mother for the remainder of her life. When she was twenty-four years old, she found another much larger marine reptile. It had a long neck and four paddle-like limbs with which it must have rowed itself through the seas. It was named Plesiosaurus (Greek for "lizard-like"). Four years later, she discovered a Pterosaur, the first flying reptile ever found in Britain.

These reptiles lived at the same time as the great land reptiles, the dinosaurs; although dinosaurs had not yet been discovered when Mary was young. While geologists did not know how long ago these reptiles lived then, they do now; they believe that it was about one hundred and sixty-five million years ago, a time known as the Jurassic period.

Mary was to find many more ichthyosaurs and plesiosaurs as well as hundreds of smaller fossils. She sold most of them to the leading geologists of the day, who also became her life-long friends. One of her customers was a king, the King of Saxony, who even came to visit her.

Henry De la Beche, after graduating as an officer, did change his mind, and he became a geologist, a famous one, for later he was knighted by Queen Victoria and became Sir Henry De la Beche.

Tray was her constant companion for years until he was killed by a rock-fall which just barely missed Mary.

The fossils that Mary collected can be seen in museums the world over—Paris, Switzerland, United States and others. Her largest and best are, of course, in the British Museum of Natural History in London.

The Misses Philpots' collection is now in Oxford Museum. Mary's fossils can also be seen in smaller museums in her own county, at Dorchester, Bridport and her own town, Lyme Regis.

Mary's grave is close to the church that she loved so well. Within the church is a stained glass window, and below it the following inscription:

"This window is sacred to the memory of Mary Anning, of this parish, who died March 9th, 1847, and is erected by the Vicar of Lyme Regis and some of the members of the Geological Society of London, in commemoration of her usefulness in furthering the science of geology, as also of her benevolence of heart and integrity of life."

BIBLIOGRAPHY

Chief Source:

Addresses of Dr. W. D. Lang, British geologist, printed in *Proceedings of the Dorset Natural History and Archaeological Society*—Vol. 60 (1938), Vol. 66 (1944), Vol. 71 (1949), Vol. 74 (1952), Vol. 76 (1954), Vol. 80 (1959), Vol. 81 (1959). (Longmans of Dorchester at the Friary Press, Dorchester)

Other Sources:

Mary Anning's letters in the British Museum, Department of Natural History, London, S.W.7.

Arkell, W. J., *The Jurassic System in Great Britain* (Oxford, Clarendon Press, 1933)

Evans, I. O., *The Observer's Book of Geology* (Frederick Warne & Co. Ltd., London, no date)

Hawkes, Jacquetta, *A Land* (Cresset Press, London, 1951)

Swinton, W. E., *The Wonderful World of Prehistoric Animals* (Garden City Books, Garden City, New York, 1961)

Swinton, W. E., *Fossil Amphibians & Reptiles* (Jarrold & Sons Ltd., Norwich, 1962)

Tourist booklet, *Lyme Regis* (Dunster's, Lyme Regis)

Turner, M. L., *Old Lyme* (Dunster's, Lyme Regis)

INDEX